Single Frequency Semiconductor Lasers

Books in the Tutorial Texts Series

Single Frequency Semiconductor Lasers
Jens Buus (1991)

An Introduction to Biological and Artificial Neural Networks for Pattern Recognition
Steven K. Rogers and Matthew Kabrisky (1991)

Laser Beam Propagation in the Atmosphere
Hugo Weichel (1990)

Infrared Fiber Optics
Paul Klocek and George H. Sigel, Jr. (1989)

Spectrally Selective Surfaces for Heating and Cooling Applications
C. G. Granqvist (1989)

Forthcoming Tutorial Texts

Aberration Theory Made Simple
Virendra N. Mahajan

Digital Image Compression: Theory and Practice
Majid Rabbani and Paul W. Jones

Single Frequency Semiconductor Lasers

Jens Buus
Plessey Research Caswell Ltd.

Donald C. O'Shea, Series Editor
Georgia Institute of Technology

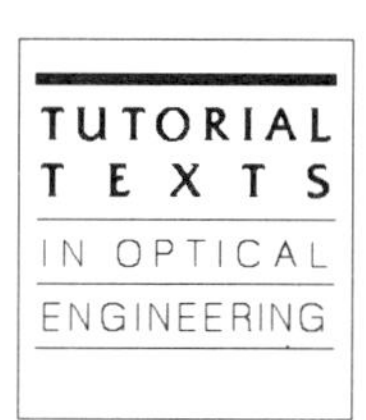

Volume TT 5

SPIE Optical Engineering Press

A Publication of SPIE—The International Society for Optical Engineering
Bellingham, Washington USA

Library of Congress Cataloging-in-Publication Data

Buus, Jens.
Single frequency semiconductor lasers / Jens Buus.
p. cm. -- (Tutorial texts in optical engineering ; TT 5)
Includes bibliographical references.
ISBN 0-8194-0535-3
1. Semiconductor lasers. I. Title. II. Series.
TA1700.B88 1991
621.36'6--dc20 90-48298
CIP

Published by
SPIE—The International Society for Optical Engineering
P.O. Box 10
Bellingham, Washington 98227-0010

Printed in the United States of America

Introduction to the Series

These Tutorial Texts provide an introduction to specific optical technologies for both professionals and students. Based on selected SPIE short courses, they are intended to be accessible to readers with a basic physics or engineering background. Each text presents the fundamental theory to build a basic understanding as well as the information necessary to give the reader practical working knowledge. The included references form an essential part of each text for the reader requiring a more in-depth study.

Many of the books in the series will be aimed to readers looking for a concise tutorial introduction to new technical fields, such as CCDs, fiber optic amplifiers, sensor fusion, computer vision, or neural networks, where there may be only limited introductory material. Still others will present topics in classical optics tailored to the interests of a specific audience such as mechanical or electrical engineers. In this respect the Tutorial Text serves the function of a textbook. With its focus on a specialized or advanced topic, the Tutorial Text may also serve as a monograph, although with a marked emphasis on fundamentals.

As the series develops, a broad spectrum of technical fields will be represented. One advantage of this series and a major factor in the planning of future titles is our ability to cover new fields as they are developing, giving people the basic knowledge necessary to understand and apply new technologies.

Donald C. O'Shea — December 1990
Georgia Institute of Technology

Contents

PREFACE

Improved fabrication technology has made it possible to fabricate reliable semiconductor lasers with low threshold current and good modulation properties. These lasers are very attractive for use as transmitters in fiber optic communication systems. There is a particular interest in single frequency semiconductor lasers for use in long-distance, high data rate systems, and as sources in fiber-based sensors.

In this text the basic physics that determine the spectral properties of semiconductor lasers will be outlined, and laser structures that give single frequency operation will be described. Modulation, noise properties, and spectral linewidth are also discussed.

The objective is to try to answer the following questions:

- Why are semiconductor lasers multimoded?
- Why do we want single mode lasers?
- How can we make single mode lasers?
- How wide is a single mode?
- Can we use semiconductor lasers in coherent systems?
- How can we make the linewidth narrower?
- Can we make even better devices?

Both fundamental concepts and practical results will be described. In some cases quite detailed theoretical analyses will be given in order to demonstrate important techniques and results. The derivation of the theoretical results are not always given in full detail, but sufficient information is provided to allow the reader to understand the principles and if necessary to fill in the remaining steps. Considerable care has been taken to ensure the correctness of all equations and figures, but, as the reader will appreciate, correctness cannot be guaranteed. The author would be grateful for information on possible mistakes as well as for comments in general.

The symbols used are listed (see Chap. 11), and for a number of parameters typical values are given. Please note that in some cases the same symbol is used for different quantities, but the meaning should be clear from the context.

A number of figures have been reproduced from the literature. All non-attributed figures originate from Plessey Research Caswell, and it is a pleasure for me to thank a number of my colleagues for providing figures: R. W. Allen, R. Ash, C. Park, P. Williams, and in particular W. Baker of the Graphics Department for much of the artwork.

I would also like to acknowledge comments to early versions of the manuscript from my colleagues. Last but not least I would like to thank the Series Editor and his Referee for providing a large number of valuable suggestions which I am sure have contributed to making the text more digestible.

Caswell
August 1990

J. Buus

CHAPTER 1. INTRODUCTION

In this introduction a very basic background on semiconductor lasers is given. For more details a number of textbooks can be recommended.[1-5] For up-to-date results the regularly published special issues of IEEE Journal of Quantum Electronics are highly recommended (the latest issues are Vol. 25, no. 6, June 1989 and Vol. 23, no. 6, June 1987).

1.1. FUNDAMENTALS

A laser is basically an oscillator working at optical frequencies. In order to get an oscillator to work, both *amplification* and *feedback* are needed. In a semiconductor laser the amplification (gain) is provided by injecting current into the active region, and the feedback is usually provided simply by reflection from the facets. Two situations must be considered: (1) steady-state operation in which the gain must be equal to the loss, and (2) dynamic behavior described by a set of rate equations. These are important concepts that form the basis for the understanding of the properties of semiconductor lasers, and we will use both the steady-state condition and the rate equations again and again.

An important property of semiconductor lasers is that they are pumped simply by passing a current through the laser structure. This is in contrast to other laser types which are usually pumped either optically or by an electrical discharge. The basic semiconductor laser structure consists of a narrow bandgap material, the active region (usually undoped) placed between two wide bandgap regions, and the confinement or passive regions, one n-doped, the other p-doped. This is known as a double heterostructure.

When this structure is forward biased the band diagram looks as shown schematically on Fig. 1.1.

Under forward bias electrons are injected into the active region, which has the band gap Eg_{act}, from the n-type passive layer, and holes are injected from the p-type passive layer. The band gap for the passive regions must be larger than the separation of the Fermi levels, otherwise the carriers would not be confined to the active region. It also turns out that the Fermi level separation has to exceed the band gap of the active region.

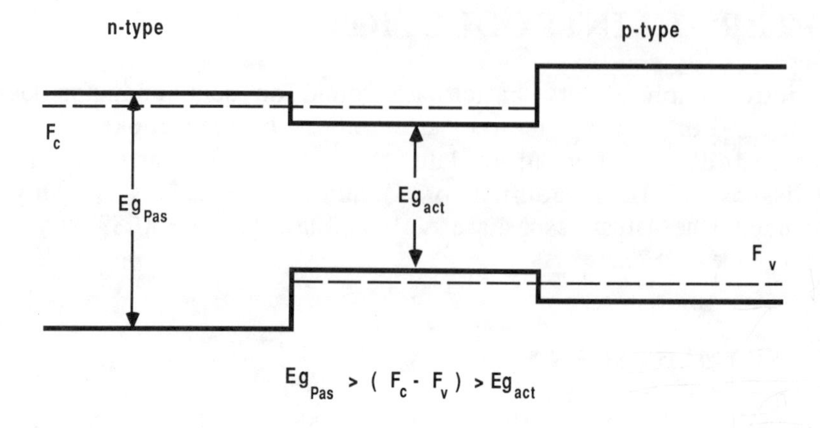

Figure 1.1. Band diagram of double heterostucture under forward bias. The Fermi levels for electrons and holes are shown by broken lines.

The carriers can recombine in the active region by several mechanisms: nonradiative recombination, bimolecular recombination, and Auger processes. The latter involves carrier scattering between different energy bands (for more details see, for example, Ref. 4). In addition to these processes stimulated recombination takes place under lasing.

The carrier recombination can be written in terms of a carrier lifetime τ_s that depends on the carrier density N. Each of the three processes mentioned above gives rise to a recombination term, and the total recombination rate is written as follows:

$$\frac{N}{\tau_s} = \frac{N}{\tau_{nr}} + BN^2 + CN^3 . \tag{1.1}$$

For later use we also introduce a differential recombination rate which is described by an effective lifetime τ_s' given by:

$$\frac{1}{\tau_s'} = \frac{1}{\tau_{nr}} + 2BN + 3CN^2 . \tag{1.2}$$

For typical carrier densities (around a few times 10^{24} m^{-3}) the terms in Eq. (1.1) are of comparable magnitude and the value of τ_s' is therefore approximately equal to $\tau_s/2$.

1.2. GAIN

The presence of carriers in the active region changes the optical properties of the material, and for sufficiently high carrier densities gain is possible. The gain is due to a population inversion created by the injection of electrons and holes into the active region. There is a large number of electrons in the conduction band and a large number of holes in the valence band.

It is possible to calculate the gain using solid-state theory. For details the reader should consult one of the textbooks on semiconductor lasers (for example, Ref. 1, which contains substantial detail). Here we are only interested in some general features of the gain. Figure 1.2 illustrates schematically the gain as a function of the photon energy with the carrier density as parameter.

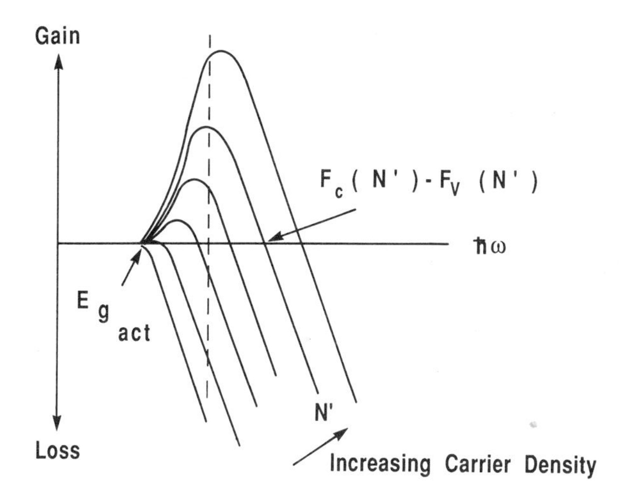

Figure 1.2. Gain as function of photon energy in a semiconductor laser (schematic).

A semiconducting material is transparent for photons of low energy, but as soon as the photon energy exceeds the band gap there is a very high absorption (or negative gain). As the carrier density increases, a photon energy range with positive gain appears. It turns out that gain is present if the photon energy is larger than the band gap of the active region but smaller than the separation between the Fermi levels.

There are some important properties of the gain that are specific to semiconductor lasers. The first is that the values of the gain are extremely high, in the range of hundreds of cm^{-1}, which is orders of magnitude greater than in any other type of laser. This is why semiconductor lasers can be made small, typically much less than 1 mm long. The second remarkable feature is that this gain curve is extremely wide, in the range of tens of nm. The reason for this is that the optical transition is between a pair of energy bands, instead of between well-defined states. There is one more interesting fact. If we look at the gain for a fixed photon energy, we can see that the gain increases with carrier density, and a very useful approximation is that it increases linearly:

$$g = a(N - N_0) . \tag{1.3}$$

This approximation will be used extensively. At this point it should be noted that the linear gain approximation is invalid for the so-called quantum well lasers. Also, effects such as spectral hole burning, which give rise to power-dependent gain, are ignored (see also Sec. 5.1).

1.3. THE ROUND-TRIP CONDITION

We have seen that the carriers are confined to the active region due to the difference in band gap. But the active region performs another very important task—it confines the photons. This is because the refractive index of the active region is higher than the refractive index of the passive regions. The active region therefore acts as an optical waveguide, as shown in Fig. 1.3.

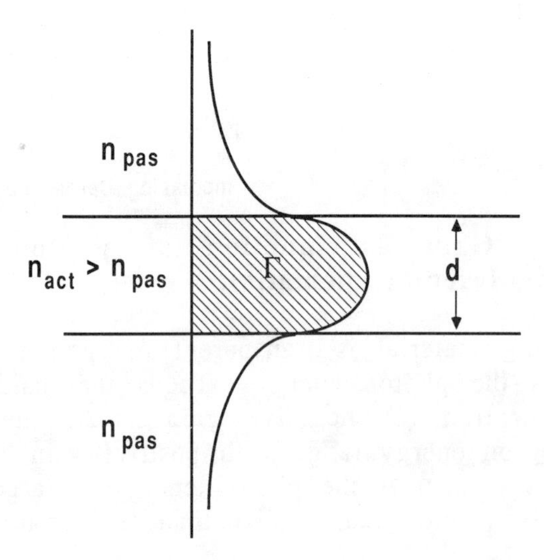

Figure 1.3. Schematic of waveguiding in a semiconductor laser, with the optical intensity distribution outlined.

Since this is a *dielectric* waveguide the optical intensity will spread into the passive regions as well. We describe this by the *confinement factor* Γ, which is the fraction of the intensity in the active region.

The value of Γ depends on the refractive index difference, the active layer thickness, and the wavelength, and it can be calculated by standard dielectric waveguide theory (e.g., Ref. 6). A very convenient, and relatively precise, approximation for Γ is given by:[7]

$$\Gamma = \frac{2v^2}{1 + 2v^2}, \tag{1.4}$$

where v is the so-called normalized frequency,[6] which, in addition to the refractive indices, also depends on the active layer thickness d and the wavelength λ:

$$v = \frac{\pi d}{\lambda}\sqrt{n_{act}^2 - n_{pas}^2}\ . \tag{1.5}$$

The net gain per unit length, assuming the passive layers to be lossless, is given by the product of the confinement factor Γ and the gain per unit length in the active region, g_{act}:

$$g_{net} = \Gamma\, g_{act}\ . \tag{1.6}$$

We now consider the steady-state condition: gain = loss. Starting at point 1 (Fig. 1.4) we make one round-trip and calculate the total amplification.

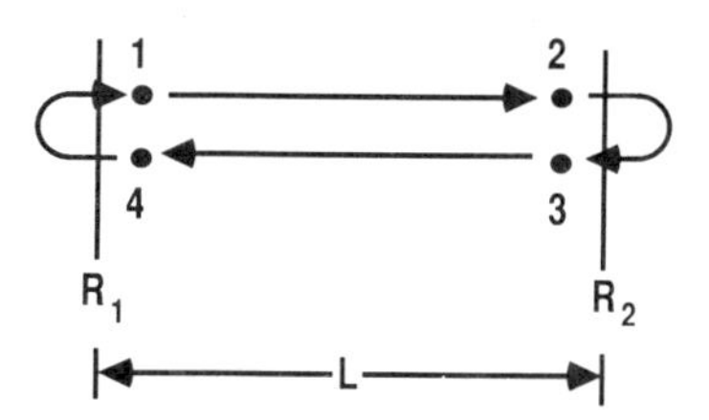

Figure 1.4. Round-trip condition.

First, from point 1 to point 2 we assume a net gain g_{net} per unit length. The distance is the laser length L, so the amplification is $\exp(g_{net}L)$. Then we hit the facet, and the fraction R_2 is reflected, and the rest of the light is emitted. On the way back we have another gain factor and another facet reflectivity. The total round-trip amplification is then:

$$A = \exp(g_{net}L)\ R_2\ \exp(g_{net}L)\ R_1\ . \tag{1.7}$$

If the round-trip gain is required to equal 1 we can find how much gain is required. This is described by putting the required net gain equal to an effective loss coefficient α_{end}, which we call the end loss (or facet loss):

$$g_{net} = \frac{1}{L}\ln\left(\frac{1}{\sqrt{R_1R_2}}\right) = \alpha_{end}\ . \tag{1.8}$$

The expression is slightly simpler if the two reflectivities are identical:

$$\alpha_{end} = \frac{1}{L}\ln\left(\frac{1}{R}\right),\ \text{for } R = R_1 = R_2\ . \tag{1.9}$$

In addition to the end loss there can of course be internal losses in the structure. In this case the net-gain-equal-to-loss condition can be written:

$$\Gamma g_{act} = \alpha_{int} + \alpha_{end} = g_{th}\ , \tag{1.10}$$

where α_{int} is the internal loss and g_{th} is called the threshold gain.

As the current (and therefore the carrier density) increases, the gain will increase. As soon as the threshold gain is reached we have satisfied the gain = loss condition and lasing starts. If the current is increased further the gain remains constant-since gain = loss is satisfied-and all extra carriers supplied by increasing the current recombine by stimulated recombination. The current where the threshold gain is reached is called the threshold current. This current marks the separation between the range where the output is dominated by spontaneous emission and the range where the output is dominated by stimulated emission, or lasing (Fig. 1.5).

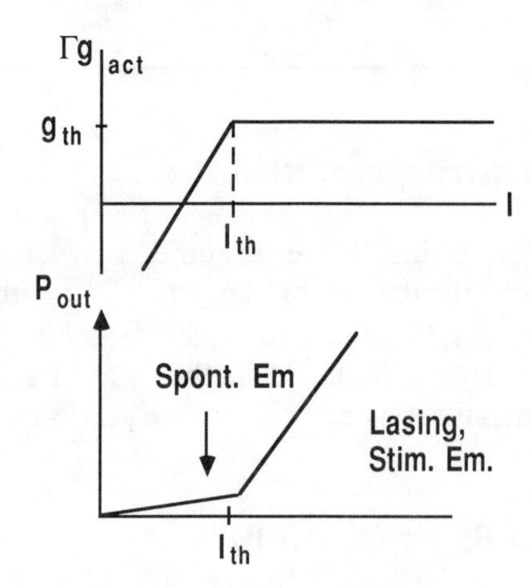

Figure 1.5. Gain and output power as functions of current.

CHAPTER 2. STEADY-STATE PROPERTIES

In this chapter we present the laser rate equations which will be used extensively in the subsequent sections. Laser properties under steady-state operation, output power, and spectrum are described.

2.1. THE RATE EQUATIONS

The behavior of a semiconductor laser is illustrated in Fig. 2.1.

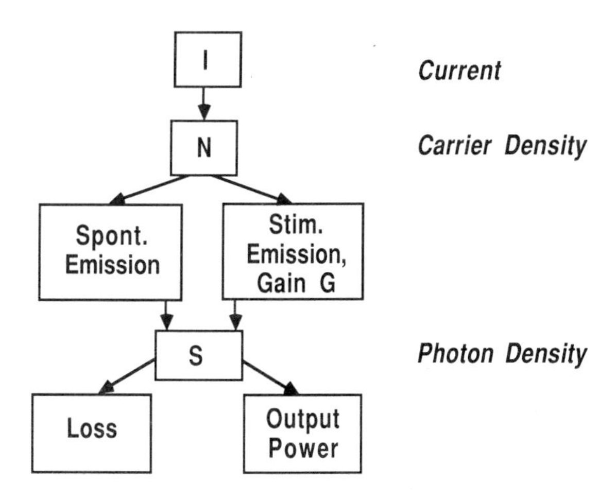

Figure 2.1. Schematic representation of a semiconductor laser.

A current I flows into the active region and supplies charge carriers to the active region. These carriers either recombine spontaneously or give rise to gain for the stimulated recombination. This leads to a photon density in the active region. Some of the photons are lost, and some of them are emitted from the laser; this gives the optical output power.

Based on this diagram, and considering the balance between the current, the carrier density, and the photon density, we can write down the two rate equations which govern the time dependence of the carrier density and the photon density.

The first equation deals with the carriers. Carriers are supplied by the current I into the active volume V. Some of the carriers recombine spontaneously, with a lifetime τ_s. Other carriers recombine by stimulated recombination, described by a gain factor G' and a photon density S. The time dependence of the carrier density is given by (e being the unit charge):

$$\frac{dN}{dt} = \frac{I}{eV} - \frac{N}{\tau_s} - G'S . \qquad (2.1)$$

The gain G' is the net gain per unit time (denoted by the prime '). This gain is related to the net gain per unit length g_{net} by:

$$G' = v_g \, g_{net} , \qquad (2.2)$$

where v_g is the group velocity of light in the laser.

The time dependence of the photon density is described by the second rate equation:

$$\frac{dS}{dt} = G'S - \alpha'S + \beta\frac{N}{\tau_s} . \qquad (2.3)$$

The first term on the right-hand side is the stimulated recombination term. We get one photon for each stimulated recombination, so this term is the same as the last term in the carrier density rate equation, but the sign is now positive. Some of the photons are lost, and this is described by a loss coefficient α'. Finally we have a term that is due to the spontaneous emission: a fraction β of the spontaneous recombination events happens to supply a photon into the lasing mode. This last term is important for the dynamic behavior; without this term and with $S = 0$ at $t = 0$, S would remain 0.

The rate equations (2.1) and (2.3) constitute a set of two coupled nonlinear differential equations. These equations form the basis for discussion of laser modulation, noise, and spectral linewidth.

2.2. OUTPUT POWER

In order to calculate how much optical power is emitted we look at the rate equation for the carriers. In steady-state the stimulated emission term must equal the pumping term minus spontaneous recombination, and the gain must be equal to the threshold gain:

$$v_g \, g_{th} \, S = \frac{I}{eV} - \frac{N}{\tau_s} = \frac{I - I_{th}}{eV} . \qquad (2.4)$$

Here we have introduced the threshold current I_{th}, which is defined as the current required to reach the carrier density where the gain equals the required threshold gain. The carrier density is related to the threshold current and is fixed above the threshold. This gives an expression for the photon density since the threshold gain equals the internal loss plus the end loss:

$$S = \frac{I - I_{th}}{eV} \frac{1}{v_g\,(\alpha_{int} + \alpha_{end})}. \qquad (2.5)$$

We are not all that interested in the photon density- what really matters is the output power. This is found as the product of the photon density, the volume (this gives the total number of photons), the photon energy, the group velocity, and the end loss [from Eq. (1.9)]. A factor 1/2 is included because we consider the output from only one facet and assume that the two reflectivities are identical:

$$P_{out} = \frac{1}{2} S\, V\, \hbar\omega\, v_g \frac{1}{L} \ln\left(\frac{1}{R}\right). \qquad (2.6)$$

By using the expression for the photon density we can write the output power in a different way:

$$P_{out} = \frac{1}{2} \frac{\frac{1}{L}\ln\left(\frac{1}{R}\right)}{\alpha_{int} + \frac{1}{L}\ln\left(\frac{1}{R}\right)} (I - I_{th}) \frac{\hbar\omega}{e}. \qquad (2.7)$$

The first part of this expression is the differential efficiency η, which is the end loss from one facet divided by the total loss. This is related to the slope of the power vs. current curve in Fig. 1.5. The second part of Eq. (2.7) is simply the power supplied above threshold. In other words, output power equals efficiency multiplied by supplied power.

2.3. THE SPECTRUM

So far we have not considered the lasing spectrum at all. We need one more condition. This is provided by writing the round-trip condition for the *field*:

$$\exp(-2j\beta L)\sqrt{R_1 R_2} = 1. \qquad (2.8)$$

Here β is the propagation constant, which contains both a real part and an imaginary part. The imaginary part is just another way of describing the gain. Note that a factor 2 appears because the gain is defined as *intensity* gain:

$$\beta = n_{eff}k + j\frac{g_{net}}{2}, \qquad k = \frac{2\pi}{\lambda}. \qquad (2.9)$$

The index n_{eff} is the effective refractive index of the waveguide structure and is given by:

$$n_{eff} = \sqrt{b\, n_{act}^2 + (1 - b)\, n_{pas}^2}\ , \qquad (2.10)$$

where b is the normalized index, which can be calculated by dielectric waveguide theory.[6] The following expression has been suggested as a simple approximation for b:[8]

$$b = 1 - \frac{1}{2v^2} \ln(1 + 2v^2)\ . \qquad (2.11)$$

When writing the round-trip condition for the *intensity*, the phase was neglected. If we write the round-trip condition for the *field*, we get a phase factor. This means that the round-trip condition splits in two: the gain condition, which we already have considered, and a phase condition, which states that the round-trip phase must equal an integer multiple of 2π. We can write this as index times length equal to an integer multiple of half the wavelength λ, which is just a resonance condition:

$$2\, n_{eff} \frac{2\pi}{\lambda} L = m\, 2\pi\ , \qquad n_{eff}\, L = m \frac{\lambda}{2}\ . \qquad (2.12)$$

From this we can find the wavelength separation between two modes:

$$\Delta\lambda = \frac{\lambda^2}{2\bar{n}L}\ . \qquad (2.13)$$

This can also be written in terms of the optical frequency ν:

$$\nu = m \frac{c}{2 n_{eff} L}\ , \qquad \Delta\nu = \frac{c}{2\bar{n}L}\ . \qquad (2.14)$$

For a laser operating at 1.55 μm with a length of about 300 μm the mode spacing is about 1 nm, or 120 GHz. This is far less than the width of the gain curve. A new term $\bar{n}$, introduced in Eqs. (2.15) and (2.16), is the group index, which we need because the refractive index is a function of the wavelength:

$$\bar{n} = n_{eff} + \nu \frac{dn_{eff}}{d\nu}\ . \qquad (2.15)$$

We can consider the solutions to the gain and phase conditions as cavity resonances. Each solution to the phase condition is a possible mode, and the required gain for each mode is found from the gain condition. In a laser with no wavelength selective elements the gain condition is independent of wavelength and the *required* gain is therefore the same for each mode. Because the mode spacing in a semiconductor laser is small compared to the width of the gain curve we have to consider more than one mode. We can still use rate equations, but now we have one photon density equation for each mode:

$$\frac{dN}{dt} = \frac{I}{eV} - \frac{N}{\tau_s} - \Sigma_i G_i'S_i, \tag{2.16}$$

$$\frac{dS_i}{dt} = G_i'S_i - \alpha'S_i + \beta\frac{N}{\tau_s}. \tag{2.17}$$

These equations describe the situation where the gain in the laser G_i' is different for the different modes, but the loss α' is the same. If we look at the steady-state situation we can can solve the photon density equations and find the photon density for each mode:

$$S_i = \frac{\beta\frac{N}{\tau_s}}{\alpha' - G_i'}. \tag{2.18}$$

This is the spontaneous emission term divided by the difference between loss and gain. Because of the presence of spontaneous emission the gain = loss condition is not exactly satisfied; instead the gain is slightly lower than the loss, and the small difference is compensated by the spontaneous emission. The wavelength dependent gain will then give a wavelength dependence of the number of photons per mode, but because the gain difference from one mode to the next is very small, we cannot expect single mode operation. This is illustrated in Fig. 2.2.

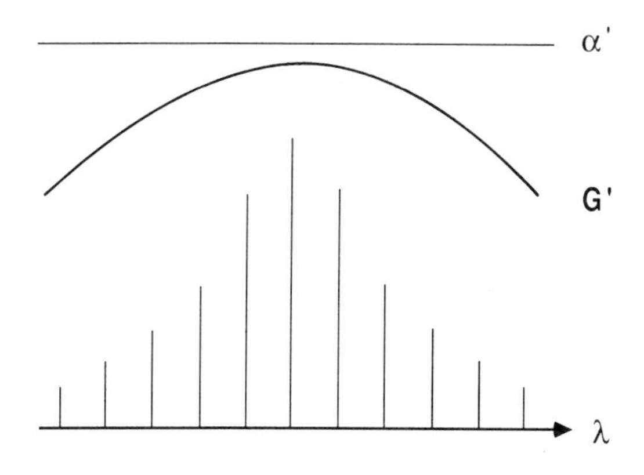

Figure 2.2. Schematic of laser spectrum.

The photon density in a mode is proportional to the reciprocal of the difference between loss and gain. When we increase the power the gain and loss curve will get closer and closer, and we should expect that the laser will have fewer and fewer modes. This happens for short wavelength GaAs lasers but not for long wavelength InGaAsP lasers. This behavior is believed to be due to some nonlinear gain mechanisms that are stronger in the long wavelength lasers. We will discuss this point in more detail later on.

We should also note that the steady-state expression for the photon density in each mode cannot be used under modulation since in this case the gain is time dependent because the carrier density varies.

A multimode spectrum has important consequences for fiber optic communication systems because the fibers are dispersive; in other words, the different modes will travel with different velocities and hence broaden the optical pulses. The condition for a communication system to work is that the pulse broadening is small compared to the duration of the pulse. The pulse broadening is given as the product of the fiber dispersion D, the fiber length L, and the laser spectral width. This broadening should be smaller than about one fourth of the pulse width ΔT (which we simply equate with 1 divided by the data rate f_B). This condition can then be written:

$$D\,L\,\delta\lambda < \frac{1}{4}\Delta T\,. \tag{2.19}$$

This condition gives the maximum value for the allowed product of the fiber length and the data rate:

$$L\,f_B < \frac{1}{4\,D\,\delta\lambda}\,. \tag{2.20}$$

We now assume that several laser modes are present, with a mode spacing of about 1 nm, giving an effective spectral width of, say, 4 nm. For standard optical fiber used at a wavelength of 1.55 μm the dispersion D is about 20 ps/(km·nm), and we find that the product of the length and data rate cannot be more than about 3 km·Gbit/s. This means that data at a rate of 1 Gbit/s can be transmitted over 3 km, and 100 Mbit/s can be transmitted over 30 km. For lower values of the fiber dispersion a higher length data rate product is of course possible, e.g., 20 km·Gbit/s for D = 3 ps/(km·nm). However, the dispersion limit is far more restrictive than the limit due to fiber loss. In order to exploit the full potential of fibers we therefore need lasers which have a much narrower spectrum.

CHAPTER 3. SINGLE MODE LASERS

In the previous chapter we saw that a multimode spectrum led to restrictions on the performance of fiber optic communication systems. It is therefore desirable to have lasers operating in a single mode. In this chapter the condition for single mode operation will be quantified and various single frequency laser structures will be presented, including external cavity lasers and distributed feedback (DFB) lasers.

3.1. SINGLE MODE CONDITION

We first need to define what we mean by a single mode laser: how single is single? Consider a case where we have one strong mode with an average power $\bar{P}_0$, and a weak side mode with an average power $\bar{P}_1$. We can then define the laser as being single moded if the ratio of the power in the strong mode to the power in the weak mode exceeds a given number. In a laser with more than one mode the power in each mode is not constant but fluctuates on a time scale in the nanosecond range. Therefore, even if there is a large ratio between the *average* powers of the strong and the weak mode, there is still a finite probability that the weak mode at a given time has more power than a certain value P'. It has been shown theoretically that the probability distribution for the power in the weak mode is exponential:[9]

$$\mathrm{prob}(P_1 > P') = \exp\left(-\frac{P'}{\bar{P}_1}\right). \tag{3.1}$$

It is obvious that $\mathrm{prob}(P_1 > 0) = 1$. If we look at the probability for the weak mode having more than one half of the average power in the strong mode we find:

$$\mathrm{prob}\left(P_1 > \frac{1}{2}\bar{P}_0\right) = \exp\left(-\frac{\bar{P}_0}{2\bar{P}_1}\right). \tag{3.2}$$

If this probability is required to be less than 10^{-9}, we find that the average power of the weak mode must be less than the average power in the strong mode divided by 42. To be on the safe side a factor of 100 is necessary.

We can look at this condition in a different way. The rate equations can be solved numerically using different values for the threshold gain of the modes. If we assume that the laser is modulated and we require a power ratio of more that a factor of 100 this will lead to a requirement for the threshold gain difference of about 5 cm^{-1}:[10]

$$\Delta g_{th} \geq 5 \text{ cm}^{-1}, \qquad \Delta g_{th} = g_{th_1} - g_{th_0}. \qquad (3.3)$$

This value is far higher than the spectral variation of the gain, and single mode operation under modulation therefore cannot be expected for ordinary lasers.

It is also possible to derive an approximate analytical expression for the required threshold gain difference (Ref. 5):

$$\Delta g_{th} = \frac{\bar{P}_0}{\bar{P}_1} \hbar\omega \, v_g \, \alpha_{end} \, (\alpha_{int} + \alpha_{end}) \frac{1}{P_{min}}. \qquad (3.4)$$

The required threshold gain difference depends on the required average power ratio, the photon energy, the group velocity, and the loss terms. The last term in the expression is the power in the "off" state; it is assumed that the laser is modulated by a digital signal. It is interesting to note that by using typical values, like those given in the list of symbols, we find again that the threshold gain difference should be about 5 cm^{-1} (using a power ratio of 100 and P_{min} = 0.05 mW). This requirement can be relaxed if the laser is biased to have a higher value of P_{min}, but this gives a lower on/off ratio.

Somehow we have to construct the laser in such a way that the dominating mode has a lower threshold, by about 5 cm^{-1}, than all other modes.

3.2. SINGLE MODE LASER STRUCTURES

A number of possible ways to construct single mode lasers have been proposed, as shown in Fig. 3.1. These can be categorized in four main groups:

1. Short lasers
2. Injection locking
3. Coupled cavities
4. Frequency selective feedback.

For a short laser the mode spacing becomes large. If the laser is so short that the mode spacing is of the same order as the width of the gain curve (tens of nm), only one mode will have a wavelength near the gain peak. However, such a laser will have to be extremely short, and consequently a very high gain is needed and/or we need very high reflectivities. Though a short laser will solve the spectral problem, devices of this type are difficult to fabricate.

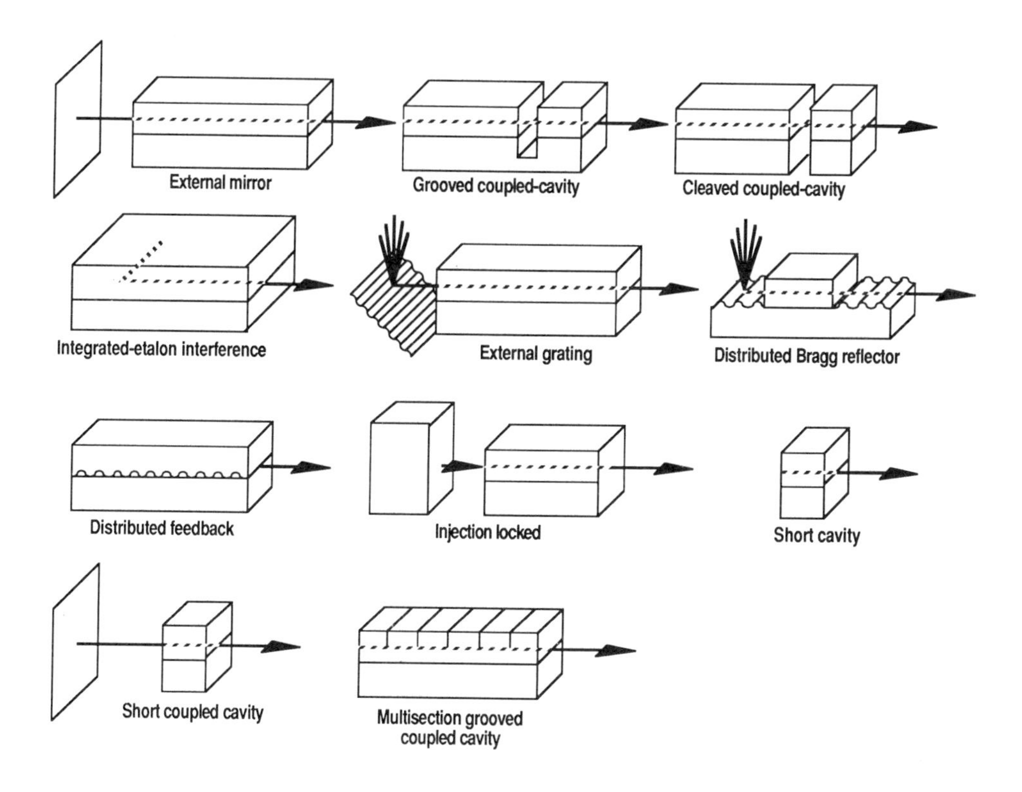

Figure 3.1. Examples of single mode lasers (Ref. 11).

Injection locking is a technique known in the context of radio frequency oscillators. A stable beam with a low power and a single mode spectrum from a "master" laser is injected into a high power "slave" laser. The problem here is that we need an isolator between the lasers in order to prevent light from the "slave" laser being coupled into the "master" laser (in this case the "slave" laser may injection lock the "master" laser). This means that it is not possible to construct a small compact device.

The two last categories of single mode lasers (coupled cavities and frequency selective feedback) are described in more detail below.

3.3. COUPLED CAVITY LASERS

Coupled cavities form a broad category that includes external cavities, groove-coupled cavities, cleave-coupled cavities, and interference lasers.

A specific example of an external cavity laser is shown in Fig. 3.2.

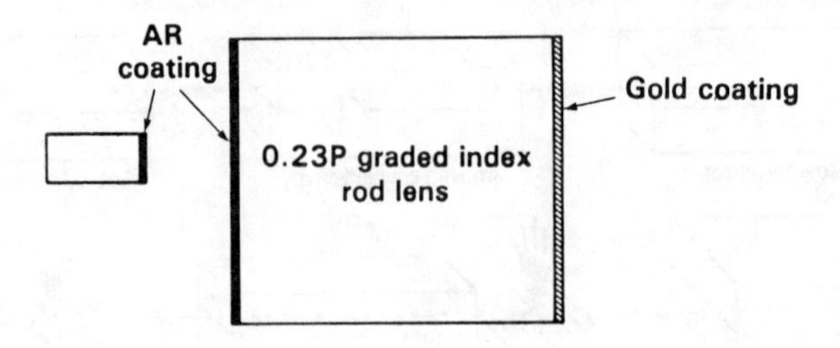

Figure 3.2. Graded index rod lens external cavity.

The advantage of this structure, which is relatively short, is that it is very compact and stable, and single mode operation is possible over a relatively wide current range.

To analyze coupled cavities we can again consider the round-trip condition (Fig. 3.3).

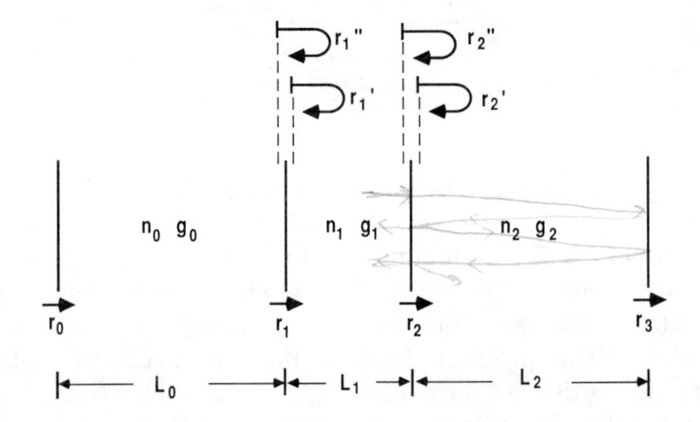

Figure 3.3. Multisection structure.

We have a number of sections, each having certain values for the refractive index and the gain. Therefore we have a reflection at each interface. We have to be very careful when we consider the signs of the reflections, because we are dealing with reflectivities for the *field*. If we consider normal incidence on the interface between region 1 with refractive index n_1 and region 2 with refractive index n_2 and light is incident from region 1, the field reflection coefficient is:

$$r = \frac{n_1 - n_2}{n_1 + n_2}. \tag{3.5}$$

Hence if light is incident from region 2 the field reflection coefficient is $-r$.

Referring to Fig. 3.3, r_0, r_1, r_2, and r_3 are defined as the reflections for a wave coming from the left. If we look from the opposite side the reflections will have the opposite sign. When we consider the part of the structure to the right of the r_2 reflector we have an effective reflectivity which also depends on the gain and index in region 2:

$$r_2' = \exp(g_2L_2 - j2kn_2L_2)\, r_3 . \qquad (3.6)$$

When we look at a point to the left of the r_2 reflector we have an effective reflectivity r_2'' that consists of the sum of r_2 and terms which arise from one, two, etc., round-trips in region 2.

$$r_2'' = r_2 + \sqrt{1 - r_2^2}\; r_2' \sqrt{1 - r_2^2}\; [\, 1 + (- r_2)r_2' + \ldots \qquad (3.7)$$

$$= r_2 + \frac{(1-r_2^2)r_2'}{1 + r_2r_2'} .$$

By using the same procedure repeatedly we can find the reflection due to the part of the structure to the right of region 0:

$$r_1' = \exp(g_1L_1 - j2kn_1L_1)\, r_2'' . \qquad (3.8)$$

$$r_1'' = r_1 + \frac{(1-r_1^2)r_1'}{1 + r_1r_1'} . \qquad (3.9)$$

We finally arrive at the round-trip condition for region 0:

$$r_1''(\lambda)\,(-r_0)\,\exp(g_0L_0 - j2kn_0L_0) = 1 . \qquad (3.10)$$

Because of the phase terms in the expressions for the reflectivities, due to the interference brought about by having several reflectors, we get a wavelength dependent reflection. In the next section we will show how the round-trip condition can be derived more systematically even for rather complicated structures.

To illustrate Eq. 3.10 we consider a very simple example with two sections only (Fig. 3.4). It is clear from the figure that we can now have a situation where the round-trip gain for one mode is higher than for all other modes. This means that the gain in the laser structure required to satisfy the round-trip condition is lower for that mode. However, the refractive index is temperature dependent and the points where the phase condition is satisfied will therefore move with changing temperature. This can lead to large mode jumps. Likewise coupled cavity lasers can usually only be modulated over a limited current range without mode jumps.

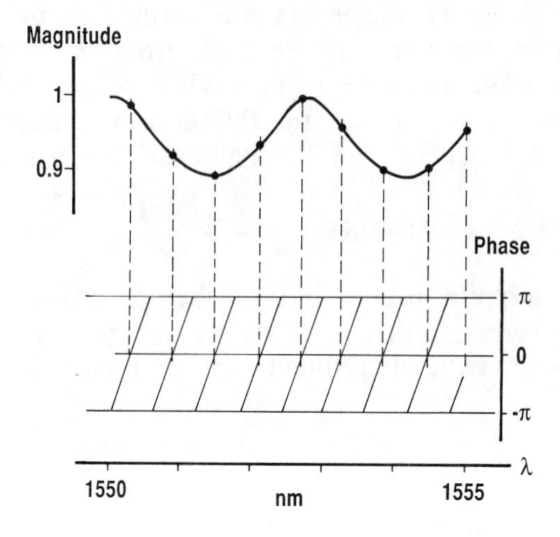

Figure 3.4. Magnitude and phase of the round-trip factor (i.e., the left-hand side of Eq. 3.10) as functions of the wavelength for a two-section structure with L_0 = 175 μm, L_1 = 135 μm, r_0 = -0.5, r_1 = 0.1, r_2 = 0.5, refractive index 3.3, and gain adjusted for a maximum round-trip gain of 1.

3.4. FREQUENCY SELECTIVE FEEDBACK

Frequency selective feedback can either be provided by an external grating or by a periodic structure in the laser. This periodic structure can either be in the form of a distributed Bragg reflector (DBR) or it can be incorporated into the active part of the laser giving distributed feedback (DFB). The fabrication of a periodic structure is illustrated in Fig. 3.5.

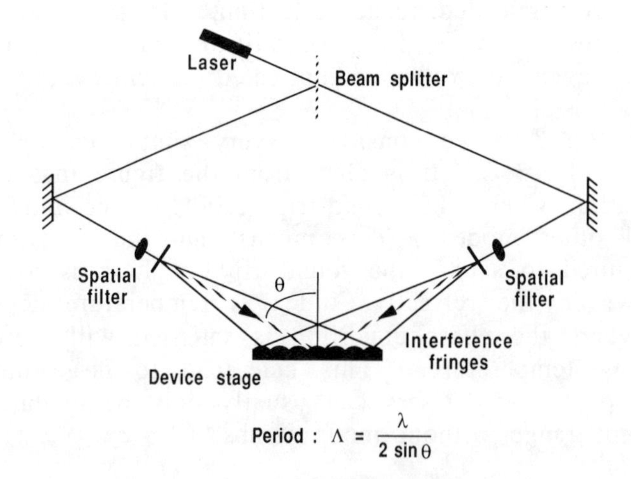

Figure 3.5. Setup for grating fabrication.

In this standard optical method, an interference pattern is formed on the surface of the material, which is coated by a photoresist. The period Λ of the interference pattern is determined by the angle θ and the wavelength of the laser used. The photoresist is developed and a grating is formed by etching. An alternative fabrication method is direct "writing" with an electron beam. This is a slower process, but it allows more freedom in the design of the grating. For lasers operating at a wavelength around 1.55 μm the period of the grating has to be around 2350 Å, or an integer multiple of this number.

Figure 3.6 shows an example of the cross section of a DFB laser with a grating which has a period of 4700 Å.

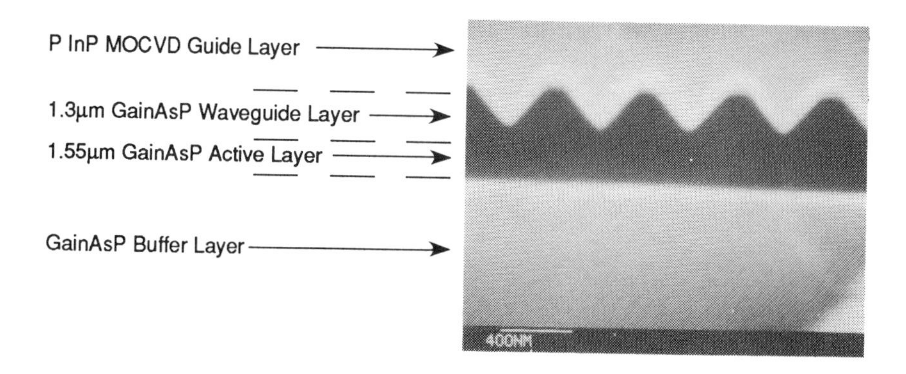

Figure 3.6. DFB laser cross section.

The grating is formed in a waveguide layer placed on top of the active layer and overgrown with material of lower refractive index, thus giving a periodic variation in the effective refractive index.

Like ordinary lasers DFB lasers need some form of lateral confinement as well. The ridge structure shown in Fig. 3.7 is attractive because the lateral optical guiding is provided by simple processing, but there is no lateral carrier confinement. The threshold current is therefore relatively high.

Structures with lateral carrier confinement are shown in Figs. 3.8 and 3.9. The complete carrier confinement leads to low threshold currents. The proton isolation shown in Fig. 3.8 reduces the parasitic capacitance and allows a higher modulation speed.

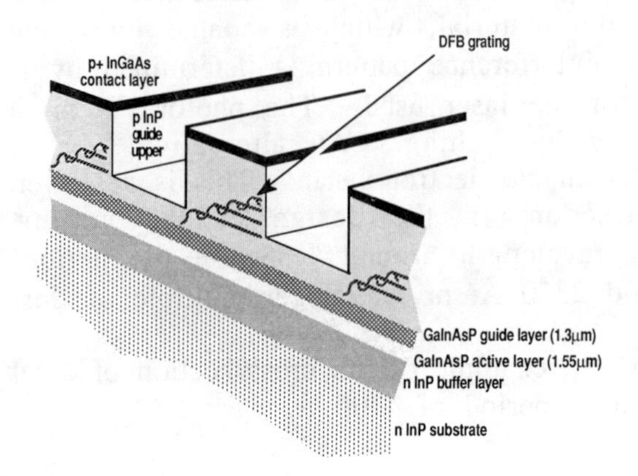

Figure 3.7. Ridge waveguide DFB.

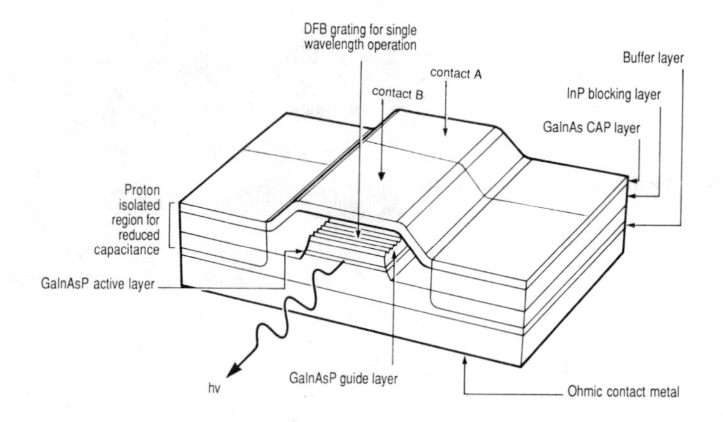

Figure 3.8. Buried ridge DFB laser.

Structures such as the buried ridge or the double channel planar buried ridge require an additional growth step to infill the structure.

These examples all show the grating formed after the growth of the active layer (grating up), but it is also possible to form the grating first (grating down)- theoretically there is no difference in performance. However, there are some practical differences.

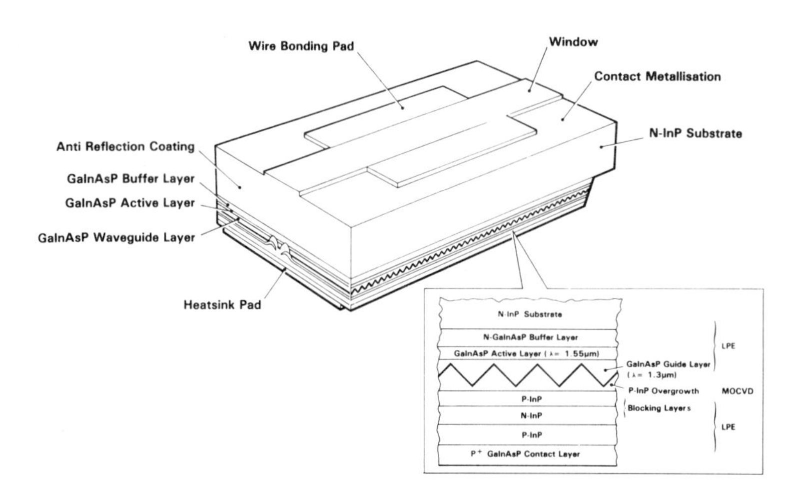

Figure 3.9. Double channel planar buried heterostructure (DCPBH) DFB laser.

If the grating is formed first, it can be formed directly in the substrate, and the complete structure can then be grown. By growing a part of the structure first and then forming the grating one more growth process is required, but it is preferable not to have to grow the active layer close to the substrate as is the case for the grating-down structures. Also it is usually easier to obtain a good regrowth on the grating, without spoiling the grating profile, if the regrown material is InP. If the grating is formed in the substrate the regrowth will have to be a quaternary InGaAsP compound.

A major advantage of the grating-up structures is that the thicknesses and compositions of the various layers, including the active layer, can be measured after the first growth. The period of the grating can then be tailored to ensure lasing at the desired wavelength. When comparing the grating-up and grating-down options one should also consider the fact that the materials growth is usually a more expensive process than grating fabrication, and the relative yields of growth and grating fabrication should be considered.

The periodic structure in a DFB or DBR laser will favor wavelengths near the Bragg wavelength. This is the wavelength where the individual reflections from the grating all add up in phase. The Bragg condition is usually expressed in terms of the propagation constant and is determined by the grating period Λ, and the order M of the grating:

$$\beta_0 = \frac{M\pi}{\Lambda}. \tag{3.11}$$

The exact wavelength (or the deviation of the propagation constant from the Bragg condition) will be determined by a phase condition. The required gain will be determined by an amplitude condition. Each set of values for the propagation constant and gain corresponds to a complex solution to the cavity resonance condition. In contrast to the simple laser structures the phase and amplitude conditions cannot be separated, and the resonance condition, which can be derived by coupled mode theory, is much more complicated. The theoretical details are discussed in Chap. 4.

The consequence of the wavelength selective reflection from the grating is that the threshold gain for modes near the Bragg wavelength will be reduced. It is therefore possible to design a laser in which one mode has a threshold gain which is lower (by more than 5 cm^{-1}) than that for all the other modes. Such lasers therefore satisfy the single mode condition.

One way of checking the performance and stability of single mode lasers is to change the temperature (Fig. 3.10).

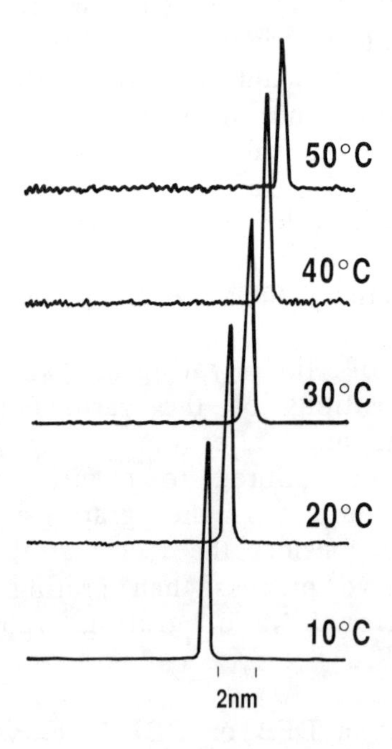

Figure 3.10. Spectrum versus temperature for DFB laser.

We see here that the laser remains single moded over a wide temperature range. The reason for the shift of the wavelength is that the refractive index and therefore the lasing wavelength increases with temperature.

A light current characteristic for a DFB laser is shown in Fig. 3.11. We see that the threshold current is quite low, with about the same value as for a good laser without a grating. The efficiency is also quite good.

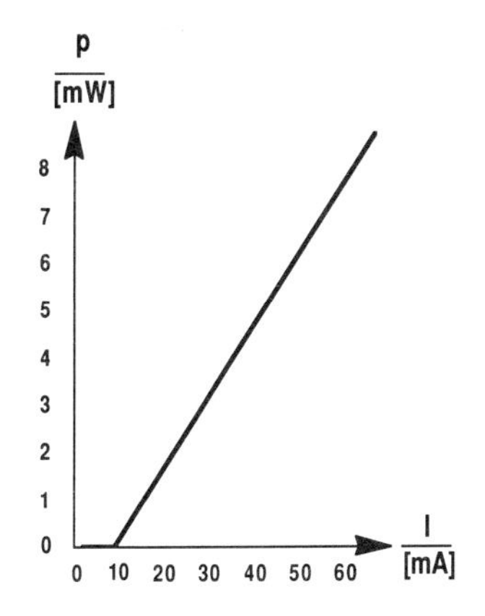

Figure 3.11. Light current characteristic for DFB laser (λ = 1.55 μm).

CHAPTER 4. PERIODIC STRUCTURES

A very high degree of wavelength selectivity can be achieved by incorporating a periodic structure into the laser. The periodic structure can either be in a layer directly above or below the active region (DFB laser) or it can form a wavelength selective reflector at the end of the laser (DBR laser). In this chapter periodic structures are analyzed by using the coupled mode equation. This forms the basis for understanding the spectral properties of DFB and DBR lasers. We also consider lasers with a combination of periodic structures and discrete facet reflections. Finally, a general method for analyzing structures with several sections, some of which may contain a periodic structure, is presented.

4.1. COUPLED MODE EQUATIONS

In order to study the optical properties of periodic structures we assume that the refractive index depends on the position in the direction of propagation:

$$n(z) = n_{eff} + \Delta n \cos(2\beta_0 z) , \quad \Delta n << n_{eff} . \tag{4.1}$$

Here β_0 is the Bragg propagation constant given by:

$$\beta_0 = \frac{M\pi}{\Lambda} = \frac{2\pi}{\lambda_0} n_{eff} , \tag{4.2}$$

where Λ is the period (or pitch) of the periodic structure and is M times a half wavelength (M is the order), and λ_0 is the Bragg wavelength. While in principle any value of M can be used, the lowest values will generally give the best performance; however, they are more difficult to fabricate because the corresponding pitch is small (Λ is around 2350 Å for $M = 1$ for a laser operating at a wavelength in the 1.55 μm range). In practice the values $M = 1$ or $M = 2$ are used. For simplicity we restrict the following discussion to the case $M = 1$.

Neglecting the lateral and transverse variations of the lasing field as well as the phase factor $\exp(-j\beta z)$ (the slowly varying envelope approximation), the wave equation reduces to:

$$\frac{\partial^2 E}{\partial z^2} + \left(n(z)\ k\right)^2 E = 0 , \tag{4.3}$$

$$\left(n(z)\,\frac{2\pi}{\lambda}\right)^2 = \beta^2 + 4\beta\,\kappa\,\cos(2\beta_0 z)\,, \tag{4.4}$$

$$\beta = n_{eff}\,\frac{2\pi}{\lambda}\,, \qquad \kappa = \frac{\pi\,\Delta n}{\lambda}\,. \tag{4.5}$$

In Eq. (4.4) a term containing Δn^2 has been neglected. It is now assumed that the propagation constant β is close to the Bragg propagation constant β_0:

$$\beta = \beta_0 + (\delta + j\alpha_0) \qquad (\delta,\, \alpha_0 << \beta_0)\,. \tag{4.6}$$

The term δ is the deviation of the real part of the propagation constant from the Bragg condition and α_0 is the gain for the *field* (note that the *intensity* gain is $2\alpha_0$).

Our next assumption is that E(z) consists of a right-propagating field plus a left-propagating field:

$$E(z) = R(z)\exp(-j\beta_0 z) + S(z)\exp(j\beta_0 z)\,. \tag{4.7}$$

Due to the presence of the Bragg propagation constant in the phase factors the functions R and S will have a comparatively weak z dependence. Inserting Eqs. (4.4) and (4.7) into Eq. (4.3), neglecting the second derivatives of R and S, and collecting terms with identical phase factors leads to the so-called coupled mode equations:

$$\frac{dR}{dz} - (\alpha_0 - j\delta)R = -j\kappa S\,, \tag{4.8}$$

$$\frac{dS}{dz} + (\alpha_0 - j\delta)S = j\kappa R\,. \tag{4.9}$$

The quantity κ is called the coupling coefficient and describes the coupling between the amplitude for the right-propagating wave R and the left-propagating wave S. In order to get a physical interpretation of the coupling coefficient we look at a wave propagating through a multilayer stack of alternating refractive index, as shown in Fig. 4.1.

The interface reflectivities are given by:

$$r_1 = \frac{(n_{eff} + \Delta n') - (n_{eff} - \Delta n')}{(n_{eff} + \Delta n') + (n_{eff} - \Delta n')} = \frac{\Delta n'}{n_{eff}}\,, \quad r_2 = -\,r_1\,. \tag{4.10}$$

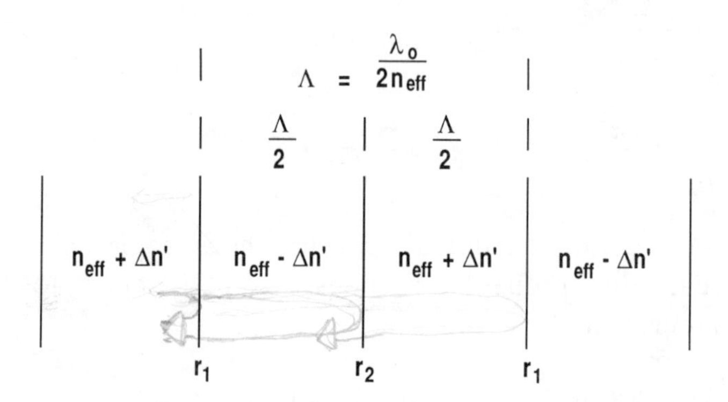

Figure 4.1. Multilayer dielectric stack, indicating period and interface reflectivities.

The signs of the interface reflectivities alternate, and since the round-trip phase change in each layer is π, all the reflected waves will add constructively. There are two reflections per period, and the coupling from the right-propagating to the left-propagating wave per length is therefore given by:

$$\kappa' = 2\frac{\Delta n'}{n_{eff}}\frac{1}{\Lambda} = 4\frac{\Delta n'}{\lambda} \ . \tag{4.11}$$

This is almost identical to the definition of κ introduced in Eq. (4.5). An extra factor of $4/\pi$ appears in Eq. (4.11) because it was derived using a rectangular variation of the index (see Fig. 4.1), instead of the cosine variation used in the derivation of the coupled mode equation. In general it is the cosine variation of the refractive index that provides the phase matching between the right- and left-propagating waves. For a given grating it is therefore necessary to calculate the Fourier expansion of the function that describes the variation of the refractive index. If this is done for the rectangular grating we get:

$$n(z) = n_{eff} + \frac{4}{\pi}\Delta n' \cos(2\beta_0 z) + \ldots \tag{4.12}$$

Using $\Delta n = \frac{4}{\pi}\Delta n'$ in (4.5) then gives Eq. (4.11).

Representing the refractive index by Eq. (4.1) means that the periodic variation has unlimited extent in the transverse and lateral directions. Figure 4.2 shows the case where the periodic index variation is restricted to a part of the structure.

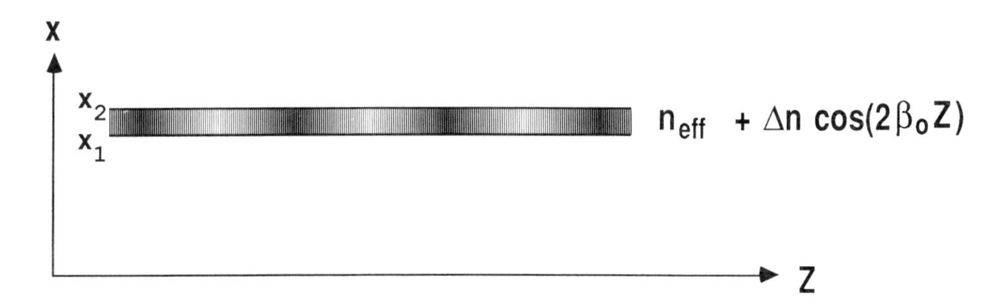

Figure 4.2. Periodic index variation, illustrated by different degrees of shading, between positions x_1 and x_2.

The periodic index variation is only "seen" by the part of the optical intensity which is present between positions x_1 and x_2 and the coupling coefficient becomes:

$$\kappa = \frac{\pi}{\lambda} \Delta n \int_{x_1}^{x_2} |E(x)|^2 \, dx \ , \tag{4.13}$$

where it is assumed that the field is normalized, i.e.,

$$\int_{-\infty}^{\infty} |E(x)|^2 \, dx = 1 \ . \tag{4.14}$$

In practice the periodic index variation is created by having a corrugation on the interface between two layers in a waveguide structure (see Fig. 3.6). For each position x' the refractive index is a step function with the value n_1 in some parts of the period and n_2 in other parts of the period. For such a structure the coupling coefficient is given by:

$$\kappa = \frac{\pi}{\lambda} \int_{x_1}^{x_2} \Delta n(x') \, |E(x')|^2 \, dx' \ . \tag{4.15}$$

In this expression $\Delta n(x')$ is the Fourier coefficient of the refractive index variation at the position x' (see Ref. 12 for specific examples).

4.2. SOLUTION OF THE COUPLED MODE EQUATIONS

The coupled mode equations (4.8) and (4.9) are a set of linear, coupled first order differential equations. The most general solution is therefore given by:

$$R(z) = r_1 \exp(\gamma z) + r_2 \exp(-\gamma z) , \qquad (4.16)$$

$$S(z) = s_1 \exp(\gamma z) + s_2 \exp(-\gamma z) . \qquad (4.17)$$

If these expressions are inserted into Eqs. (4.8) and (4.9) we find that in order to have nontrivial solutions we must have:

$$\gamma^2 = \kappa^2 + (\alpha_0 - j\delta)^2 . \qquad (4.18)$$

We look at a structure of length L, extending from $z = -L/2$ to $z = L/2$, with zero reflectivity at the ends. Since there are no fields propagating *into* the periodic structure we have:

$$R(-L/2) = S(L/2) = 0 , \qquad (4.19)$$

giving

$$r_1/r_2 = s_2/s_1 = -\exp(\gamma L) . \qquad (4.20)$$

Since the structure is symmetric with respect to z the field must be either symmetric or antisymmetric; hence,

$$r_1 = \pm s_2, \quad r_2 = \pm s_1 . \qquad (4.21)$$

The field distribution is then given by:

$$R(z) = \sinh\big(\gamma(z + L/2)\big) , \qquad (4.22)$$

$$S(z) = \pm \sinh\big(\gamma(z - L/2)\big) . \qquad (4.23)$$

In addition, by inserting Eqs. (4.22) and (4.23) into Eq. (4.8) and evaluating the resulting expression at $z=-L/2$ we have the following characteristic equation:

$$\kappa = \frac{\pm j\gamma}{\sinh(\gamma L)} , \qquad (4.24)$$

which can also be written, using Eq. (4.18):

$$(\alpha_0 - j\delta) = \gamma \coth(\gamma L) . \qquad (4.25)$$

For a given value of the coupling coefficient κ (and hence κL) Eq. 4.24 (or Eq. 4.25) constitutes an eigenvalue equation that determine the possible values of the complex number $(\alpha_0 - j\delta)$. Each solution corresponds to a mode with the threshold (field) gain given by α_0 and a wavelength determined by the value of δ. This corresponds to the amplitude and phase conditions for a Fabry-Pérot laser, but now the modes have different values for the required gain (i.e., the value of α_0). Lasing will take place in the mode that has the lowest required gain, and single mode operation is possible if the modal gain difference is sufficiently large.

The required intensity gain for a DFB laser (i.e. $2\alpha_0$) corresponds to the end loss (α_{end}) for a Fabry-Pérot laser (see Sec. 1.3). The laser theory discussed in Chaps. 5 and 6 therefore also applies to DFB lasers, provided α_{end} is replaced by $2\alpha_0$.

For the case of strong coupling ($\kappa >> \alpha_0$) the following approximate results are found for the lowest order mode (the mode with the smallest value of δ):

$$\delta L \approx \kappa L, \quad \alpha_0 L \approx \left(\frac{\pi}{\kappa L}\right)^2 . \tag{4.26}$$

These results can be derived by splitting Eq. (4.24) or Eq. (4.25) into real and imaginary parts and using series expansions.

Since Eq. (4.24), or Eq. (4.25), is a complex, trancendental equation it cannot be solved analytically in general; instead numerical methods must be used. Figure 4.3 shows the results.

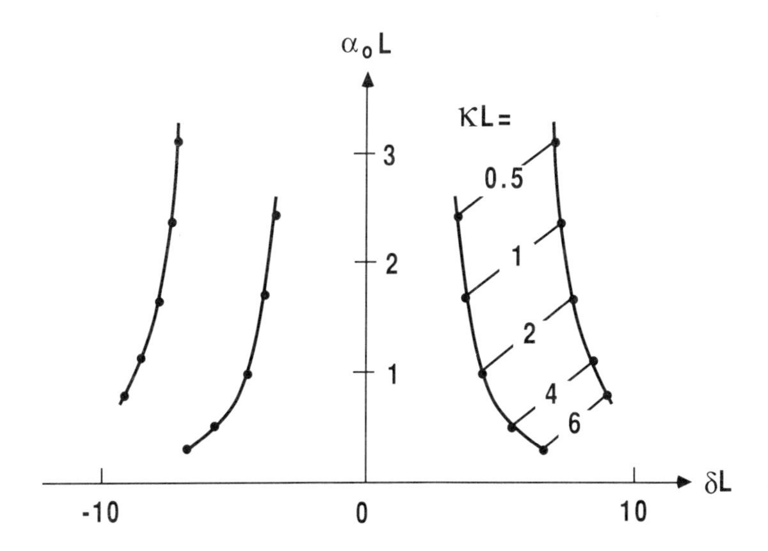

Figure 4.3. Several solutions to the characteristic equation for a periodic structure are shown as functions of κL in the (δL,$\alpha_0 L$) plane. The thin lines indicate solutions found for the same value of κL.

For a Fabry-Pérot laser the figure corresponding to Fig. 4.3 would consist of points with the same value of $\alpha_0 L$ (the actual value being dependent on the facet reflectivity), equally spaced in δL. From Fig. 4.3 we notice that there is no mode at $\delta = 0$ (corresponding to the wavelength $\lambda_0 = 2\, n_{eff}\, \Lambda$). Instead two modes exist with the same value of α_0, separated by a "stop band" that has the width $2\kappa L$ for large values of κL. The existence of two modes with the same required gain means that single mode operation will not be possible.

In order to understand why there is no solution at $\delta = 0$ we consider a simple rectangular grating (Fig. 4.4).

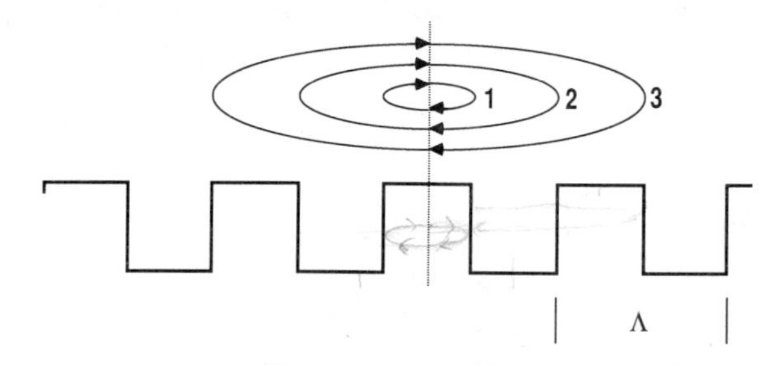

Figure 4.4. Periodic structure with rectangular grating shape, showing the individual interface reflections.

As in Fig. 4.1, the signs of the interface reflections alternate, and the phase shift for each half period is π. All the reflected waves will therefore add in phase. However, if we look at the total round-trip phase for each wave we find that it is $(2p+1)\pi$, p being an integer, and not a multiple of 2π as required. Consequently the wavelength must deviate from λ_0.

There is a simple way (at least in theory) to obtain a single mode that has $\delta = 0$ and a lower required gain than all other modes. The solution is to insert an extra phase shift in the center of the periodic structure. If the length of the phase shift section is $\Lambda/2 = \lambda_0/4n_{eff}$ (phase shift $\pi/2$), corresponding to an extra round-trip phase shift of π, all the reflected waves will add in phase and the total round-trip phase will be a multiple of 2π (Fig. 4.5).

The eigenvalue equation (corresponding to Eq. 4.24 or Eq. 4.25) for the $\pi/2$ phase-shifted grating can be derived by using the method described in Sec. 4.5.

$$\gamma \coth\left(\frac{\gamma L}{2}\right) - (\alpha_0 - j\delta) = \pm\kappa \tag{4.27}$$

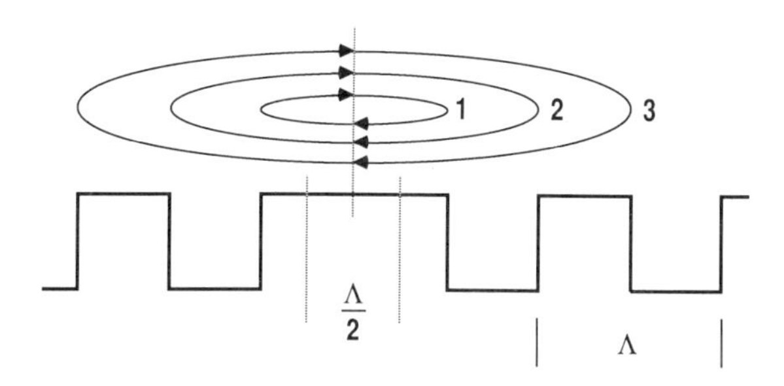

Figure 4.5. Rectangular grating with a $\pi/2$ phase shift in the center.

The solutions to Eq. 4.27 are shown in Fig. 4.6, there is now a single dominating mode at $\delta = 0$.

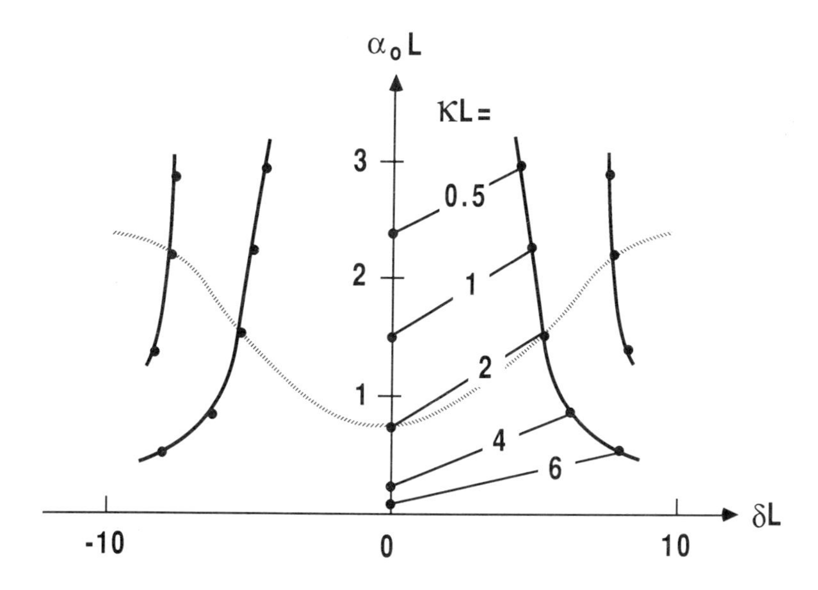

Figure 4.6. Solutions to the characteristic equation for a $\pi/2$ phase-shifted grating shown as functions of κL (the broken line shows the solution for $\kappa L = 2$ as a function of the magnitude of the phase shift).

4.3. DISTRIBUTED BRAGG REFLECTOR

A periodic structure can be interpreted as a distributed reflector with a wavelength dependent reflection. In a DBR laser a periodic structure is placed at one or both ends of the laser. To analyze the properties of a DBR laser we consider a periodic structure of length L (Fig. 4.7).

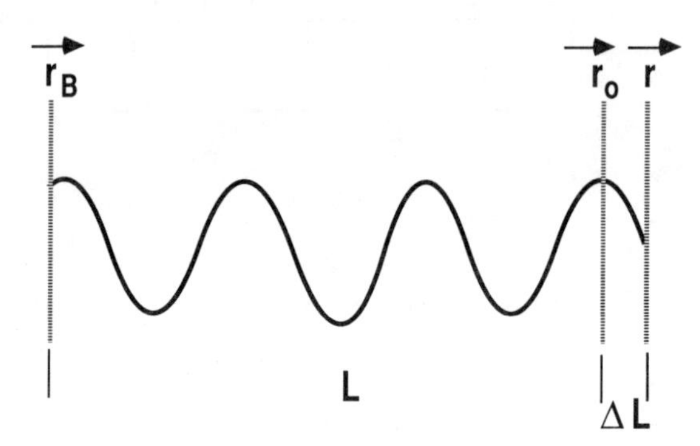

Figure 4.7. Periodic structure, left-hand side reflection r_B. The length L is equal to an integer number of periods, and the phase of the structure with respect to the right-hand side is $\Theta = 2\pi n_{eff}\Delta L/\lambda$.

The reflectivity r_B can be found using the elements of the transfer matrix given in Sec. 4.5. For the case where the right-hand side reflection (r_0 in Fig. 4.7) can be neglected the reflectivity is:

$$r_B = \frac{-j\kappa \sinh(\gamma L)}{\gamma \cosh(\gamma L) - (\alpha_0 - j\delta) \sinh(\gamma L)} . \tag{4.28}$$

As usual α_0 is the field gain (α_0 is negative if there is loss), κ is the coupling coefficient, and γ is given by Eq. (4.18).

At the Bragg wavelength, given by $\delta = 0$, γ is real and the phase of r_B is $-\pi/2$. This gives an alternative explanation for the properties of the phase shifted grating since each "half" of this structure has a reflection with a $-\pi/2$ phase shift; combined with a π round-trip phase due to the phase-shift of the grating, this gives a total round-trip phase shift of 0, thus satisfying the phase condition.

Including the right-hand side reflection r_0 gives:

$$r_B = \frac{\big(-j\kappa + (\alpha_0 - j\delta)r_0\big) \sinh(\gamma L) + \gamma r_0 \cosh(\gamma L)}{\gamma \cosh(\gamma L) - \big((\alpha_0 - j\delta) - j\kappa r_0\big) \sinh(\gamma L)} . \tag{4.29}$$

The reflection r_B now depends on the phase Θ of the right-hand side reflection since:

$$r_0 = r \exp(-2j\Theta) , \tag{4.30}$$

where r is the facet reflectivity.

4.4. DFB LASERS WITH REFLECTING FACETS

The analysis in Sec. 4.2 was restricted to DFB lasers with nonreflecting facets. The case with reflecting facets is more complicated since the discrete reflections from the facets interfere with the distributed reflections from the periodic structure. A characteristic equation for calculation of values of α_0 and δ for the various modes can be formed by multiplying the expression for the right-hand side reflectivity [Eq. (4.29)] with a similar expression for the left-hand side reflectivity and requiring the resulting expression to be equal to 1. In Fig. 4.8 a result is shown for the case of one reflecting facet.

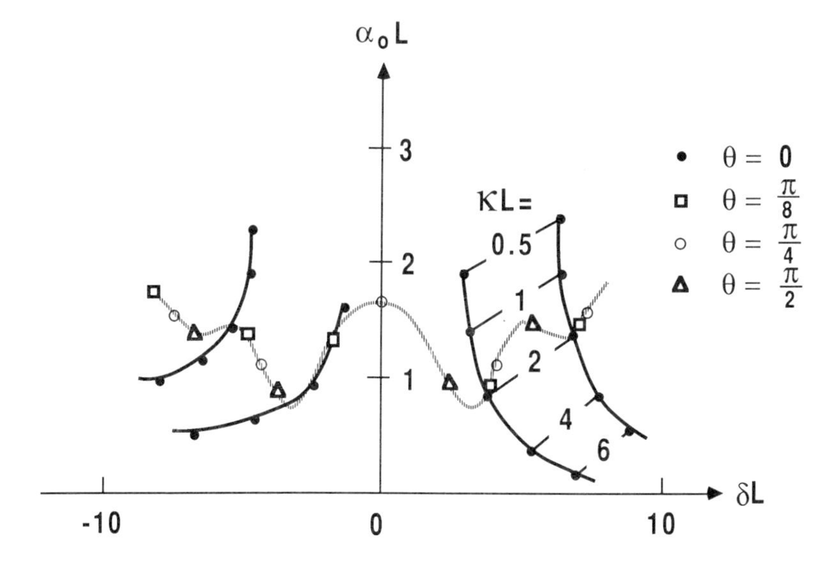

Figure 4.8. Modes represented as points in the ($\delta L, \alpha_0 L$) plane, for a periodic structure with one reflecting facet (r = 0.565, Θ = 0) as functions of κL. The broken curve shows the solution for κL = 2 as a function of Θ; modes for selected values of Θ are marked on the curve.

The presence of a reflecting facet lifts the mode degeneracy seen in Fig. 4.3, and the mode selectivity $\Delta\alpha_0 L$ (defined as the difference in $\alpha_0 L$ for the two modes with the lowest values of the required gain) becomes a function of the facet phase, i.e., the position of the facet with respect to the periodic structure. The degeneracy is still present for some phase values (e.g., $\pi/4$ in Fig. 4.8). For some values of the phase there is good mode selectivity (e.g., $\pi/8$), while for other values $\Delta\alpha_0 L$ is small (e.g., 0 and $\pi/2$).

In the fabrication of DFB lasers it is not possible to control the facet position. Consequently some lasers will have good mode selectivity while others will operate in two modes. We can define the *yield* of single mode lasers as the fraction of cases where the mode selectivity $\Delta\alpha_0 L$ is larger than a certain value. For example, an (intensity) gain difference of over 5 cm^{-1} (see Chap. 3) for a laser length of 300 μm requires $\Delta\alpha_0 L > 0.08$. For given values of κL and of the magnitudes of the facet reflectivities, a number of different values for the facet phases can be analyzed, and the number of cases where the mode selectivity is sufficiently high can be counted. Examples are shown in Figs. 4.9 and 4.10.

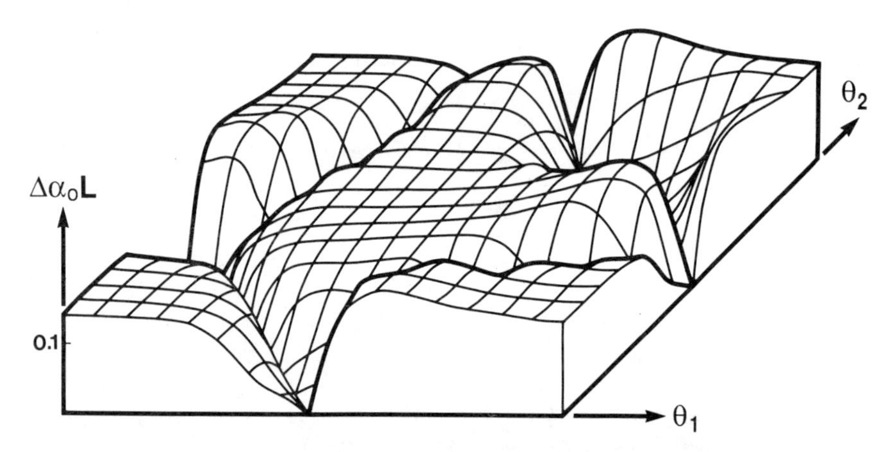

Figure 4.9. Mode selectivity for a structure with two reflecting facets, where r = 0.565 for both facets and κL = 3. $\Delta\alpha_0 L$ is shown as a function of the two facet phases.

The options for fabricating single mode DFB lasers are as follows: A $\pi/2$ phase shifted grating is used and both facets are antireflection coated; in this case the laser will always have a single dominating mode at λ_0 (if only one facet is coated not all lasers will have good mode selectivity). Alternatively it is much simpler to fabricate lasers without a phase shift and with simple cleaved facets. The price for the simpler fabrication is a reduced yield of single mode lasers.

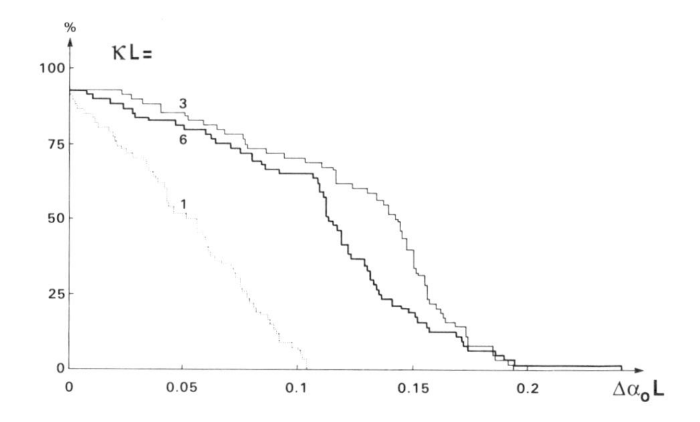

Figure 4.10. Percentage of cases where $\Delta\alpha_0 L$ exceeds the value given on the axis for a structure with two reflecting facets, both with r = 0.565. Sixteen values of each facet phase (i.e., 256 cases) were considered. The results are shown for 3 values of κL (Ref. 13).

4.5. ANALYSIS OF MULTISECTION LASERS

It is obvious that the analysis of lasers with several sections, possibly also containing periodic structures, is considerably more complicated than the analysis of a simple Fabry-Pérot laser. One convenient method is the "transfer matrix" analysis, which is outlined below.

We consider a structure consisting of a number of homogeneous sections. For each section we look at the right-propagating field E_r and the left-propagating field E_s at the left-hand side and on the right-hand sides of the section (Fig. 4.11).

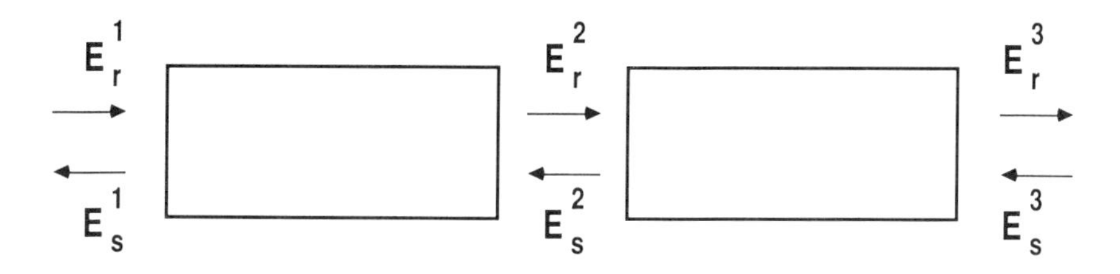

Figure 4.11. Right- and left-propagating fields in a multisection structure.

The fields on the left-hand side and on the right-hand side of section i are related by a transfer matrix $\bar{\bar{F}}_i$:

$$\begin{Bmatrix} E_r^{i+1} \\ E_s^{i+1} \end{Bmatrix} = \begin{Bmatrix} A & B \\ C & D \end{Bmatrix} \begin{Bmatrix} E_r^{i} \\ E_s^{i} \end{Bmatrix} = \bar{\bar{F}}_i \begin{Bmatrix} E_r^{i} \\ E_s^{i} \end{Bmatrix} . \tag{4.31}$$

This formulation is somewhat similar to the S matrix description of microwave components. Indeed the elements of the F matrix can be shown to be related to the elements in the S matrix. The transfer matrix for the complete structure is found by matrix multiplication:

$$\bar{\bar{F}} = \bar{\bar{F}}_N \bar{\bar{F}}_{N-1} \ldots \bar{\bar{F}}_2 \bar{\bar{F}}_1 , \tag{4.32}$$

$$\begin{Bmatrix} E_r^{N+1} \\ E_s^{N+1} \end{Bmatrix} = \bar{\bar{F}} \begin{Bmatrix} E_r^{1} \\ E_s^{1} \end{Bmatrix} . \tag{4.33}$$

For a laser there are no incoming fields $\left(E_s^{N+1} = E_r^1 = 0\right)$, but there are outgoing fields $\left(E_r^{N+1} \neq 0 \text{ and } E_s^1 \neq 0\right)$. This gives the lasing condition:

$$F_{22} = 0 . \tag{4.34}$$

We now look at the transfer matrix for some simple cases. A *field* reflection r (seen from the left-hand side) is described by:

$$\bar{\bar{F}}(r) = \begin{Bmatrix} \dfrac{1}{1-r} & \dfrac{-r}{1-r} \\ \dfrac{-r}{1-r} & \dfrac{1}{1-r} \end{Bmatrix} \tag{4.35}$$

Conversely the reflection seen from the left-hand side can be written $r = -F_{21}/F_{22}$.

For a homogeneous section of length L with a propagation constant β (note that β is complex if there is gain or loss) we have:

$$\bar{\bar{F}}(L) = \begin{Bmatrix} \exp(-j\beta L) & 0 \\ 0 & \exp(j\beta L) \end{Bmatrix} . \tag{4.36}$$

The complete transfer matrix for a Fabry-Pérot laser of length L, left-hand side field reflectivity r_1 and right-hand side field reflectivity r_2, is then:

$$\bar{\bar{F}} = \begin{Bmatrix} \dfrac{1}{1-r_2} & \dfrac{-r_2}{1-r_2} \\ \dfrac{-r_2}{1-r_2} & \dfrac{1}{1-r_2} \end{Bmatrix} \begin{Bmatrix} \exp(-j\beta L) & 0 \\ 0 & \exp(j\beta L) \end{Bmatrix} \begin{Bmatrix} \dfrac{1}{1-r_1} & \dfrac{-r_1}{1-r_1} \\ \dfrac{-r_1}{1-r_1} & \dfrac{1}{1-r_1} \end{Bmatrix} . \tag{4.37}$$

The lasing condition $F_{22} = 0$ gives:

$$\exp(-2j\beta L)\,(-r_1 r_2) = 1\,, \tag{4.38}$$

which is recognized as the round-trip condition Eq. (2.8) (except that we are now using field reflectivities instead of power reflectivities).

For a full laser analysis the round-trip condition must be combined with a rate equation for the carrier density. This will determine the value of the intensity since the amount of stimulated recombination must be such that the carrier density gives the correct gain (the threshold condition). This will satisfy the real part of Eq. (4.38). The fact that the real part of the refractive index depends on the carrier density will determine the lasing wavelength through the imaginary part of Eq. (4.38).

The transfer matrix description is particularly convenient because it is possible to define a transfer matrix for a section with a periodic structure, such as the grating in a DFB or DBR laser. We use the parameters $\delta = \beta - \beta_0$ (δ is the deviation of the real part of the propagation constant β from the Bragg propagation constant β_0), α_0 (field gain coefficient), and κ (coupling coefficient for the periodic structure). The transfer matrix can be derived from the general solution to the coupled mode equations:

$$R(z) = \left(\cosh(\gamma z) + \frac{\alpha_0 - j\delta}{\gamma}\sinh(\gamma z)\right) R(0) - \frac{j\kappa}{\gamma}\sinh(\gamma z)\, S(0)\,, \tag{4.39}$$

$$S(z) = \frac{j\kappa}{\gamma}\sinh(\gamma z)\, R(0) + \left(\cosh(\gamma z) - \frac{\alpha_0 - j\delta}{\gamma}\sinh(\gamma z)\right) S(0)\,. \tag{4.40}$$

The transfer matrix for a periodic structure of length L is then:

$$\overline{\overline{F}}_{DFB} = \begin{Bmatrix} \cosh(\gamma L) + \dfrac{\alpha_0 - j\delta}{\gamma}\sinh(\gamma L) & -\dfrac{j\kappa}{\gamma}\sinh(\gamma L) \\ \dfrac{j\kappa}{\gamma}\sinh(\gamma L) & \cosh(\gamma L) - \dfrac{\alpha_0 - j\delta}{\gamma}\sinh(\gamma L) \end{Bmatrix}. \tag{4.41}$$

In this expression the length is assumed to be an integer multiple of the grating period, and the parameter γ is defined by Eq. (4.18).

The lasing condition $F_{22} = 0$ now gives:

$$(\alpha_0 - j\delta) = \gamma \coth \gamma L . \tag{4.42}$$

This result is well known from the theory for DFB lasers (Eq. 4.24, Sec. 4.2). The characteristic equation for a DFB laser with a phase shift [Eq. (4.27)] can be found by combining two matrices of the same type as Eq. (4.41), with a phase shift given by Eq. (4.36).

A periodic structure, like the one shown in Fig. 4.12–where the length is not an integer multiple of the grating period, and where the facets are reflecting–can be analyzed by combining the elementary matrices defined above.

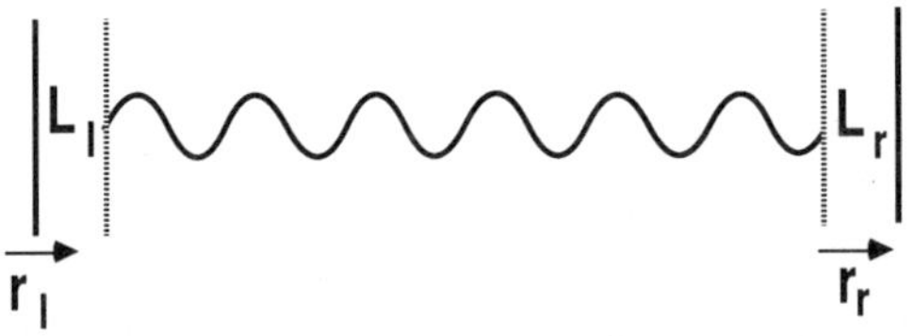

Figure 4.12. Periodic structure with reflecting facets.

The transfer matrix for this structure is:

$$\overline{\overline{F}} = \overline{\overline{F}}(r_r)\ \overline{\overline{F}}(l_r)\ \overline{\overline{F}}_{DFB}\ \overline{\overline{F}}(l_l)\ \overline{\overline{F}}(r_l) . \tag{4.43}$$

Here $\overline{\overline{F}}(r_r)$ and $\overline{\overline{F}}(r_l)$ are found from Eq. (4.35) and $\overline{\overline{F}}(l_r)$ $\overline{\overline{F}}(l_l)$ from Eq. (4.36).

The transfer matrix analysis can be applied to multisection structures, including active and passive waveguides, active and passive periodic structures, and reflectors. For a multisection structure in general, the gain (imaginary part of β) and phase (real part of β) for each section is related to the carrier density in that region. This in turn depends on the current to that section and on the optical intensity. The lasing condition is found from the transfer matrix for the complete structure.

CHAPTER 5. MODULATION AND NOISE

In this chapter we will use the rate equation to discuss modulation and noise properties of semiconductor lasers. First we will introduce some modifications of the rate equations in order to make them more suitable for this analysis. Small signal amplitude modulation is then investigated by linearizing the rate equations. The dependence of the real part of the refractive index gives rise to frequency modulation and chirping. Finally, laser intensity noise will be discussed.

5.1. MODIFICATIONS OF THE RATE EQUATIONS

The modulation properties of semiconductor lasers can be investigated by using the rate equations for the carrier density and for the photon density. These equations, however, are nonlinear and cannot in general be solved analytically. We first rewrite them in a more convenient form, recalling the expressions for the gain from Chaps. 1 and 2, and using the approximation that the gain increases linearly with carrier density. Again the prime symbol ' denotes gain per unit time. From Eqs. (1.3), (1.6), and (2.2) we have:

$$G' = v_g\, g_{net} = v_g\, \Gamma\, g_{act} = v_g\, \Gamma\, a\, (N - N_0)\,. \tag{5.1}$$

The parameter N_0 is called the transparency carrier density (the carrier density where the gain is 0), and the parameter a is called the gain slope:

$$a = \frac{dg_{act}}{dN}\,. \tag{5.2}$$

The loss is given by:

$$\alpha' = v_g\, (\alpha_{int} + \alpha_{end}) = v_g\, g_{th}\,. \tag{5.3}$$

It turns out that the modulation properties cannot be accurately accounted for without modifying the gain expression. By comparing theory and experiments (see, for example, Ref. 14) it is found that g_{act} should be replaced by an expression that depends on the photon density:

$$g_{act} \rightarrow g_{act}\,(1 - \varepsilon S)\,. \tag{5.4}$$

This may seem a bit surprising since semiconductor lasers are usually regarded as being homogeneously broadened (i.e., when the carrier density is changed by changing the amount of stimulated emission it will affect the gain at all parts of the spectrum). Recent investigations (e.g., Refs. 15 and 16) have shown that the broadening is not completely homogeneous; some degree of spectral hole burning is present. The simplest way to account for this effect is to use a formula like Eq. (5.4). It should be noted that this effect is more pronounced in long wavelength InP-based lasers than in GaAs lasers. This also explains why conventional long wavelength lasers are usually multimoded. As a final modification we also rewrite the amount of spontaneous emission into the lasing mode by replacing β (N/τ_s) in Eq. (2.3) with R/V where the spontaneous emission rate R is:

$$R = v_g \, g_{th} \, n_{sp} \, . \tag{5.5}$$

This means that the spontaneous emission is contributing by n_{sp} photons per unit volume. Note that in Eq. (5.5) it has been assumed that the carrier density is fixed at N_{th}. This is a good approximation above threshold as small variations in R have no influence on the dynamic properties. The factor n_{sp} arises because the gain is negative for a carrier density of 0, as illustrated in Fig. 5.1.

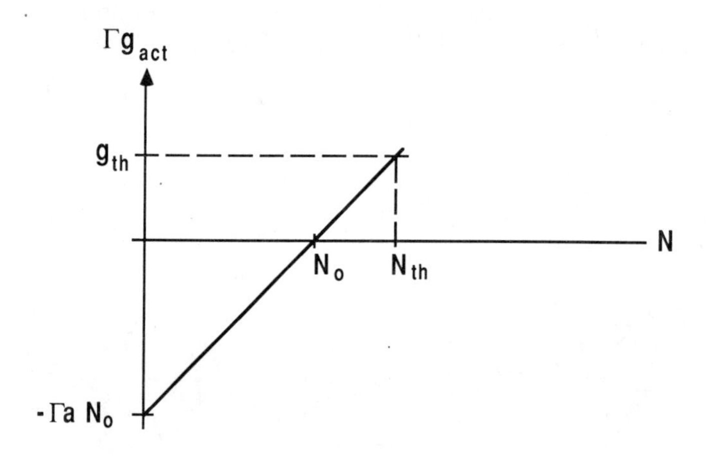

Figure 5.1. Gain as a function of carrier density (at the lasing wavelength).

From Fig. 5.1 we derive the definition:

$$n_{sp} = \frac{g_{th} + \Gamma \, a \, N_0}{g_{th}} = \frac{N_{th}}{N_{th} - N_0} \, . \tag{5.6}$$

Here n_{sp} is called the spontaneous emission factor, or the inversion parameter, and has a value between 2 and 3 for semiconductor lasers. For many other laser systems it is 1.

With these changes the rate equations (2.1) and (2.3) become:

$$\frac{dN}{dt} = \frac{I}{eV} - \frac{N}{\tau_s} - v_g \, \Gamma \, a \, (N-N_0) \, (1-\varepsilon S) \, S \, , \tag{5.7}$$

$$\frac{dS}{dt} = v_g \left(\Gamma \, a \, (N-N_0) \, (1-\varepsilon S) - (\alpha_{int} + \alpha_{end}) \right) S + \frac{R}{V} . \tag{5.8}$$

5.2. AMPLITUDE MODULATION

We now linearize the rate equations, meaning that we restrict the investigation to small signal modulation, where the current consists of a constant part and a (small) modulation term with modulation frequency ω_m:

$$I = \bar{I} + \Delta I \, \exp(j\omega_m t) \tag{5.9}$$

For small signal modulation we can write both the carrier density and the photon density as a constant part and a time-dependent part:

$$N = \bar{N} + \Delta N \, \exp(j\omega_m t) \, , \tag{5.10}$$

$$S = \bar{S} + \Delta S \, \exp(j\omega_m t) \, . \tag{5.11}$$

For operation above the lasing threshold we can make the following approximations:

$$\bar{N} \approx N_{th} \, , \qquad v_g \, \Gamma \, a \, (\bar{N}-N_0) \approx G_{th}' \, . \tag{5.12}$$

The gain depends on both carrier density and photon density:

$$G'(\bar{N},\bar{S}) = v_g \, \Gamma \, a \, (\bar{N}-N_0) - v_g \, \Gamma \, a \, (\bar{N}-N_0) \, \varepsilon \, \bar{S} \tag{5.13}$$

$$= \frac{\delta G'}{\delta N} (\bar{N}-N_0) + \frac{\delta G'}{\delta S} \bar{S} \, , \qquad \frac{\delta G'}{\delta S} \approx - G_{th}' \, \varepsilon \, . \tag{5.14}$$

We now insert Eqs. (5.9)–(5.11) into the rate equations, use Eqs. (5.12)–(5.14), eliminate the steady-state solution, and neglect higher order terms. Also we have to remember that the carrier lifetime τ_s depends on the carrier density [see Eq. (1.1)]. Therefore the differential lifetime τ_s' appears. The result is:

$$j \, \omega_m \, \Delta N = \frac{\Delta I}{eV} - \frac{\Delta N}{\tau_s'} - \frac{\delta G'}{\delta N} \bar{S} \, \Delta N - G_{th}' \, \Delta S \, , \tag{5.15}$$

$$j \, \omega_m \, \Delta S = \left(\frac{\delta G'}{\delta N} \Delta N + \frac{\delta G'}{\delta S} \Delta S \right) \bar{S} - \frac{R}{V\bar{S}} \Delta S \, . \tag{5.16}$$

The two terms in Eq. (5.16) that contain ΔS are damping terms, and we define the damping parameter as:

$$\gamma = G_{th}' \varepsilon \bar{S} + \frac{R}{V\bar{S}} . \tag{5.17}$$

We can now solve Eqs. (5.15) and (5.16). For the carrier density modulation we find:

$$\frac{\Delta N}{\Delta I} =$$

$$\frac{1}{eV} \frac{\gamma + j\omega_m}{\frac{\delta G'}{\delta N}\bar{S} G_{th}' + \gamma\underline{\left(\frac{1}{\tau_s'} + \frac{\delta G'}{\delta N}\bar{S}\right)} + j\omega_m\left(\gamma + \underline{\frac{1}{\tau_s'} + \frac{\delta G'}{\delta N}\bar{S}}\right) + (j\omega_m)^2} . \tag{5.18}$$

The terms underlined in Eq. (5.18) turn out to be small compared with the other terms and can be neglected. This gives:

$$\frac{\Delta N}{\Delta I} = \frac{\gamma}{e\,V\,\frac{\delta G'}{\delta N}\,\bar{S}\,G_{th}'} \left(1 + j\frac{\omega_m}{\gamma}\right) H(\omega_m) . \tag{5.19}$$

The function $H(\omega_m)$ is given by

$$H(\omega_m) = \frac{1}{1 + \frac{j\omega_m \gamma}{\omega_r^2} + \left(\frac{j\omega_m}{\omega_r}\right)^2} . \tag{5.20}$$

Since the linearized rate equations form a second order system we have a resonance frequency given by:

$$f_r = \frac{\omega_r}{2\pi} = \frac{1}{2\pi}\sqrt{\frac{\delta G'}{\delta N}\,\bar{S}\,G_{th}'} = \frac{1}{2\pi}\sqrt{\frac{v_g \Gamma a}{eV}}\,\sqrt{I - I_{th}} . \tag{5.21}$$

This means that the resonance frequency increases with the square root of the output power and depends on the gain slope a. Usually the first term on the right-hand side of Eq. (5.17) dominates, and we then have the following relation between the damping parameter and the resonance frequency:

$$\gamma = K f_r^2 , \quad K = \frac{4\pi^2 \varepsilon}{v_g \Gamma a} . \tag{5.22}$$

It has been argued that K, which for long wavelength lasers is found to have a value around 0.3 ns, is a material constant.[17]

The solution for the photon density modulation is given by:

$$\frac{\Delta S}{\Delta I} = \frac{1}{eVG_{th}'} H(\omega_m) . \tag{5.23}$$

This can also be written in terms of a power modulation, introducing the laser efficiency η:

$$\frac{\Delta P_{out}}{\Delta I} = \eta \frac{\hbar\omega}{e} H(\omega_m) . \tag{5.24}$$

Since $H(0) = 1$ this gives the dc response we expected. The transfer function $H(\omega_m)$ is shown in Fig. 5.2.

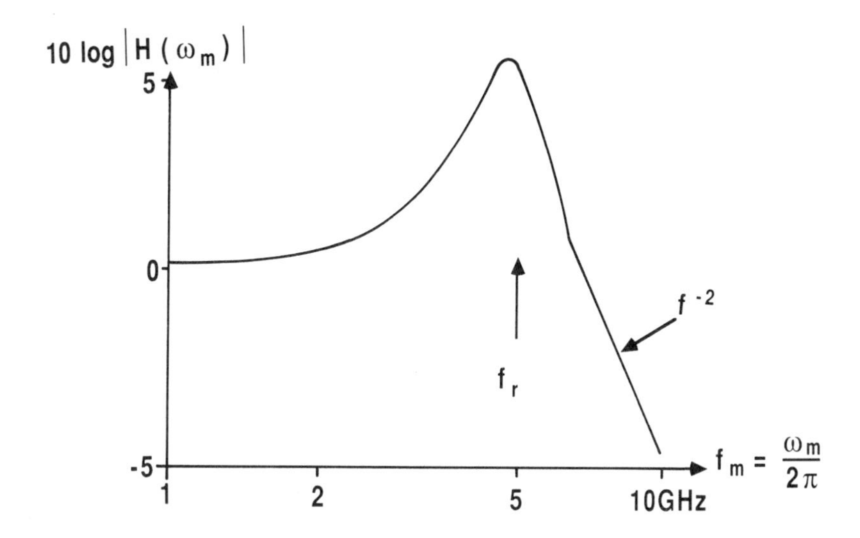

Figure 5.2. Amplitude modulation transfer function as a function of the modulation frequency.

In Fig. 5.2 we can clearly see the resonant behavior (note that the peak appears for a modulation frequency slightly below f_r). The height of the resonance peak depends on the damping parameter γ, which according to Eq. (5.17) depends on the nonlinear gain parameter ε. With $\varepsilon=0$ the calculated resonance peak would be much higher and not agree with experimental measurements of amplitude modulation.

When the modulation response of a laser is considered we must keep in mind that the transfer function $H(\omega_m)$ refers to the intrinsic optical response only. Electrical parasitics such as parallel capacitance and series resistance as well as package parasitics will give rise to an "electrical" transfer function $H_{el}(\omega_m)$. As the simplest possible description this can be represented by the response of a parallel RC circuit with a cutoff frequency ω_{el}:

$$H_{el}(\omega_m) = \frac{1}{1 + \left(\frac{\omega_m}{\omega_{el}}\right)^2} . \tag{5.25}$$

The total modulation response is the product of $H(\omega_m)$ and $H_{el}(\omega_m)$ (see Fig. 5.3).

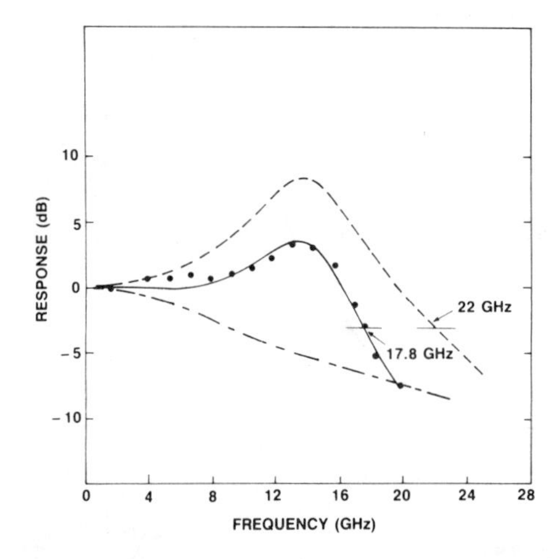

Figure 5.3. Modulation response of a laser. The upper curve shows $H(f_m)$, the lower curve shows $H_{el}(f_m)$, and the full line is the product. The experimental points are shown as dots. For this laser f_r = 14.4 GHz and f_{el} = 9.7 GHz (Ref. 18).

The example shown in Fig. 5.3 is for a 1.3 μm InGaAsP laser and is one of the highest modulation bandwidths reported. Up to 17 GHz bandwidth has been reported for a 1.55 μm DFB laser.[19]

5.3. FREQUENCY MODULATION

In a semiconductor laser the real part of the refractive index (and therefore the optical frequency) depends on the carrier density. Modulation of the laser current will therefore also give rise to frequency modulation. We will first look at the reason why the refractive index depends on the carrier density. For most laser systems the gain is due to a transition between two discrete states, giving a narrow, symmetrical gain curve. Gain is related to the imaginary part of the refractive index, and presence of gain (or loss) will, according to the Kramers-Kronig relations, lead to a change in the real part of the refractive index. This is illustrated in Fig. 5.4.

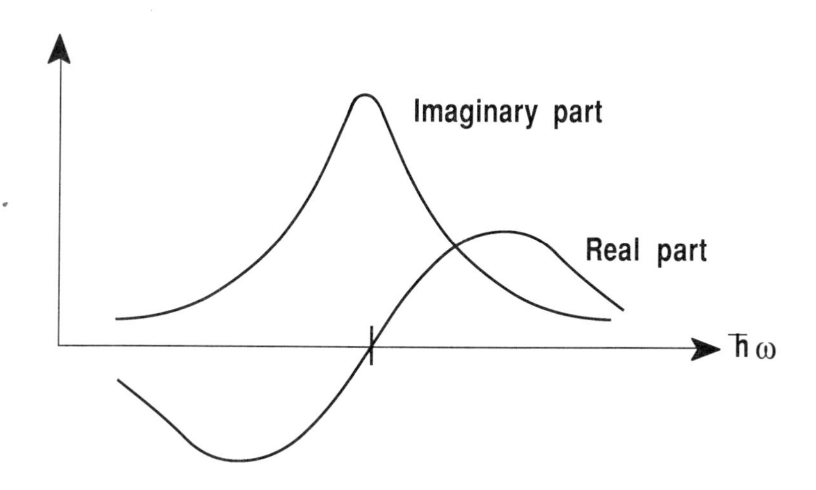

Figure 5.4. Variation of the real and imaginary parts of the refractive index around a laser transition.

For a symmetrical gain curve there is no change in the real part of the refractive index at the photon energy where the gain has its maximum. For a semiconductor laser the gain curve is not symmetrical leading to an asymmetrical variation in the real part of the index. The consequence of this is that there will be a significant change in the real part of the refractive index around the lasing transition (see Fig. 5.5).

We describe the change in the real part of the refractive index by the parameter α, which is defined as the ratio between the change in the real part of the propagation constant and the change in the imaginary part of the propagation constant when the carrier density is changed (for more details see Ref. 20).

$$\alpha = \frac{\frac{d\mathrm{Re}\{\beta\}}{dN}}{\frac{d\mathrm{Im}\{\beta\}}{dN}} = 2k\frac{\frac{dn_{eff}}{dN}}{\frac{dg_{net}}{dN}} = 2k\frac{\Gamma\frac{dn_{act}}{dN}}{\Gamma\frac{dg_{act}}{dN}} = 2k\frac{a'}{a}, \qquad (5.26)$$

$$a' = \frac{dn_{act}}{dN}, \quad a = \frac{dg_{act}}{dN}, \quad \alpha < 0 \text{ since } a' < 0 \text{ and } a > 0. \qquad (5.27)$$

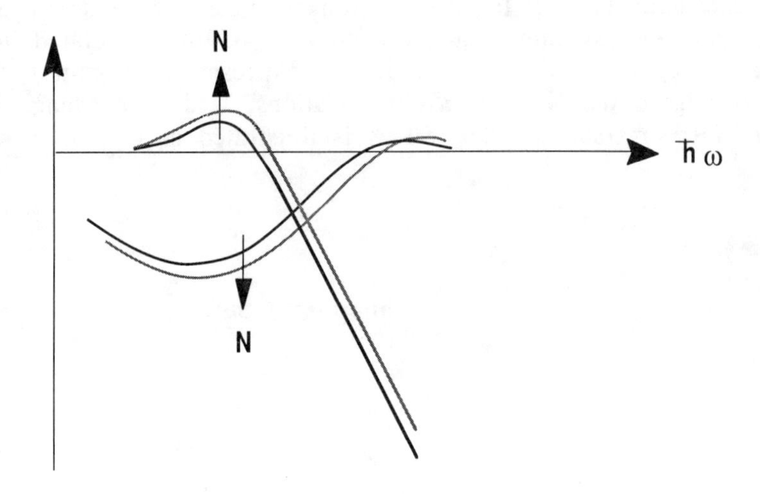

Figure 5.5. Variation of the gain (upper curve) and real part of the refractive index (lower curve) for a semiconductor laser. Changes in the curves due to an increase of the carrier density are also shown.

In Eq. (5.26) we have assumed that the waveguiding properties of the laser are not changed when the refractive index is changed. The change in the effective refractive index n_{eff} equals the change in the refractive index of the active layer multiplied by the confinement factor, and the change in the net gain equals the change in the gain of the active layer multiplied by the confinement factor.

In this section we will see that the α parameter is responsible for frequency modulation. In Sec. 5.4 we will see that it gives rise to frequency chirping, and later on we will see that it also has consequences for laser linewidth and for the behavior of external cavity lasers. For long wavelength InP-based semiconductor lasers the value of α is about -5.

We can now write an expression for the change of the optical frequency due to a change of the carrier density:

$$\Delta f = -c\frac{\Delta\lambda}{\lambda^2} = -\frac{c}{\lambda}\frac{\Gamma\Delta n_{act}}{\bar{n}} = -\frac{c}{\lambda}\frac{\Gamma\alpha\Delta g_{act}}{2k\bar{n}} = -\frac{\alpha}{4\pi}v_g\,\Gamma\, a\,\Delta N\,. \tag{5.28}$$

An increasing carrier density leads to a decreasing refractive index (and increasing gain), giving a decreasing wavelength and an increasing optical frequency (remember that α is negative).

Alternatively we can write Eq. (5.28) as a rate equation for the optical phase:

$$\frac{d\phi}{dt} = -\frac{\alpha}{2} v_g \Gamma a (N - \bar{N}) + \text{constant}. \tag{5.29}$$

From Eq. (5.19) we know the relation between current modulation and carrier density modulation. Using this in Eq. (5.28) gives a relation between the optical frequency modulation and the current modulation:

$$\frac{\Delta f}{\Delta I} = -\frac{\alpha}{4\pi}\frac{\delta G'}{\delta N}\frac{\Delta N}{\Delta I} = -\frac{\alpha}{4\pi}\frac{\gamma}{eV\bar{S}G_{th}'}\left(1 + j\frac{\omega_m}{\gamma}\right)H(\omega_m). \tag{5.30}$$

Again we have a resonant behavior, as shown in Fig. 5.6.

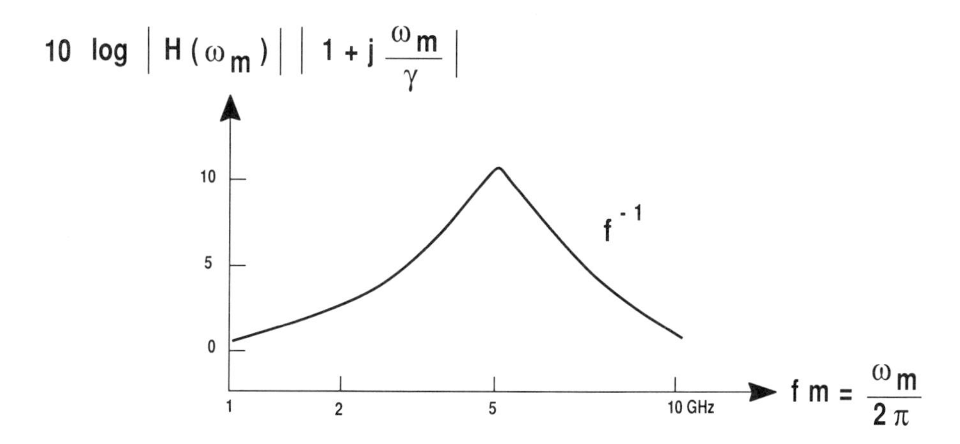

Figure 5.6. Frequency modulation transfer function as a function of the modulation frequency.

For low frequencies we have:

$$\frac{\Delta f}{\Delta I} = -\frac{\alpha}{4\pi}\frac{\gamma}{eV\bar{S}G_{th}'} \tag{5.31}$$

$$= -\frac{\alpha}{4\pi}\frac{\varepsilon}{eV} \quad \text{for} \quad G_{th}'\,\varepsilon\,\bar{S} >> \frac{R}{V\bar{S}}. \tag{5.32}$$

In the absence of the nonlinear gain parameter ε, the frequency modulation would be very weak at low modulation frequencies, but including ε gives a better agreement with experimental results. In the expressions above we have neglected the fact that current modulation gives rise to temperature modulation, and the refractive index depends on the temperature. Assuming that the thermal properties can be described by a simple thermal cutoff frequency we have:

$$\frac{\Delta f}{\Delta I} = -\frac{\alpha}{4\pi}\frac{\gamma}{eV\bar{S}G_{th}'} + \frac{\frac{df}{dT}\frac{dT}{dI}}{1 + j\frac{\omega_m}{\omega_{th}}} . \tag{5.33}$$

Here ω_{th} is in the MHz range. The low frequency (sub-MHz) modulation of the optical frequency has a value of about -1 GHz/mA. An experimental result for the optical frequency modulation is shown in Fig. 5.7.

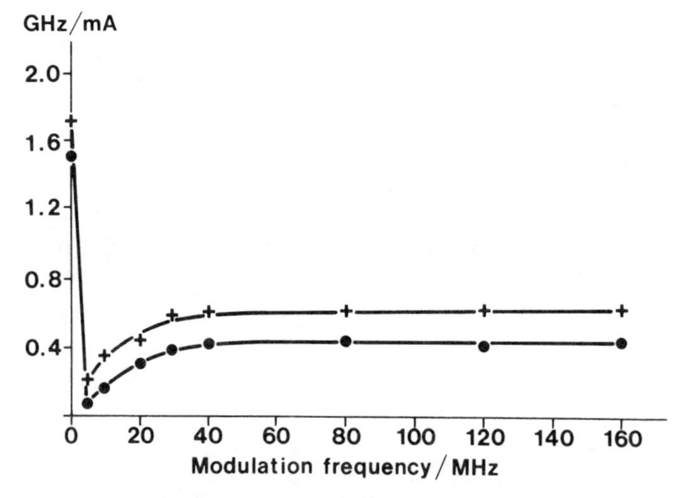

Figure 5.7. Measurement of the magnitude of the optical frequency modulation.

From Eq. (5.30) we can find the optical frequency modulation (FM) index $2\pi|\Delta f|/\omega_m$, and from Eq. (5.23) we can find the amplitude modulation (AM) index $\Delta S/\bar{S}$. If we take the ratio of these quantities we find:

$$\frac{\text{FM index}}{\text{AM index}} = \frac{|\alpha|}{2}\frac{\gamma}{\omega_m}\left|1 + j\frac{\omega_m}{\gamma}\right| . \tag{5.34}$$

The magnitude of this expression is:

$$\frac{|\alpha|}{2}\sqrt{1 + \left(\frac{\gamma}{\omega_m}\right)^2} . \tag{5.35}$$

An experimental result for this quantity is shown in Fig. 5.8.

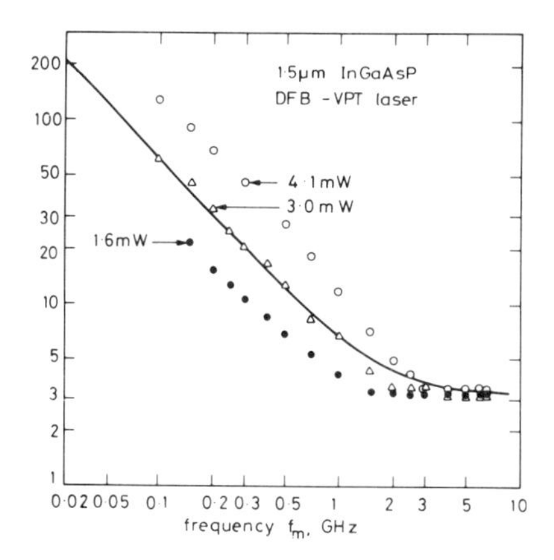

Figure 5.8. Measured magnitude of the FM/AM ratio shown as function of the modulation frequency (Ref. 21).

The measurement of the frequency dependence of the FM/AM ratio makes it possible to determine both the damping parameter γ and the α parameter. For small values of γ, the curve would be flat down to low frequencies. Again the presence of nonlinear gain seems to be necessary in order to explain the experimental results.

5.4. CHIRPING

With a single mode laser the spectral width is orders of magnitude smaller than for a multimode laser, less than 0.001 nm rather than several nm. So now we might think that we could forget completely about fiber dispersion, but that is not the case. We have a very fundamental limitation: the time bandwidth product of a pulse. If we want to transmit information the laser must be modulated. This means that we create sidebands. If we have a Gaussian pulse we know that the temporal width of the pulse multiplied by the spectral width is 0.44; this is known as a transform-limited pulse:

$$\Delta T \, \Delta f = \frac{2 \ln 2}{\pi} = 0.44 \, . \tag{5.36}$$

We note here that this time bandwidth limit follows from the well-known time energy uncertainty relation $\Delta T \, \Delta E \geq \hbar/2$. The numerical factor in Eq. (5.33) arises because we consider the full width at half maximum rather than the RMS width. If we now look at the dispersion condition, which is that the dispersion gives a pulse broadening that is less than one fourth of the inverse of the data rate, we get the following condition:

$$D\,L\,\frac{\lambda^2}{c}\,\Delta f < \frac{1}{4\,f_B}\,. \tag{5.37}$$

Taking the temporal pulse width to be approximately the inverse of the data rate and the spectral width to be approximately equal to the data rate (this corresponds to $\Delta T\ \Delta f\ =\ 1$), we get the following limitation:

$$L\,f_B{}^2 < \frac{c}{4\,\lambda^2\,D}\,. \tag{5.38}$$

This is a restriction on the value of the product of transmission distance and the *square* of the data rate. This is because the modulation-induced broadening exceeds the spectral width of the laser. (We remember that for a multimode laser the dispersion gave a limit on the product of the transmission distance and the data rate.) Inserting numbers indicates that a very long distance should be possible, but unfortunately this result is not quite right.

In order to derive a more correct result we first look at the solution to the rate equations when a laser is modulated. We start with a current pulse and get an optical pulse, as shown in Fig. 5.9.

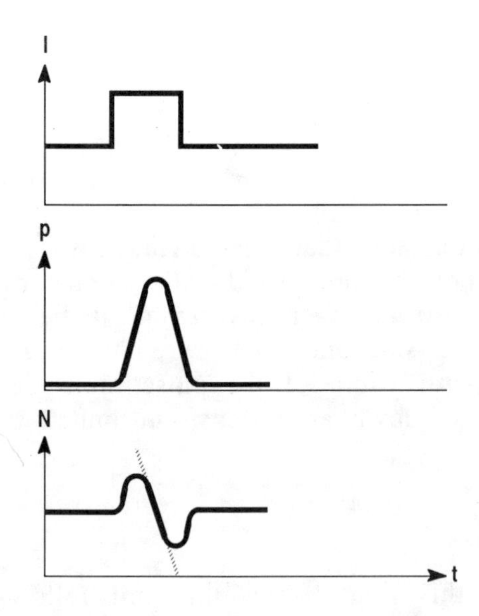

Figure 5.9. Variation in power P and carrier density N, following a current pulse.

We see that the carrier density varies during the pulse. This is explained as follows. When the current is increased we start to increase the carrier density, which means that the gain will increase.

When the gain is high the photon density will increase rapidly, but eventually the photon density gets so high that because of the stimulated recombination term in the carrier rate equation the carrier density will start to fall. When it falls the gain will fall, and when gain minus loss becomes negative the photon density will start to decrease.

For a semiconductor laser the refractive index depends on the carrier density, and a decrease of the carrier density will lead to an increase in the refractive index. An increasing refractive index means an increasing wavelength (or decreasing frequency) since the resonance condition has to be maintained. The shift to longer wavelengths during a pulse is also called "redshift." The gradual variation of the frequency is called "chirping."

Figure 5.10 shows a measurement of a time resolved spectrum during a pulse (the spectral width is limited by the spectrometer resolution).

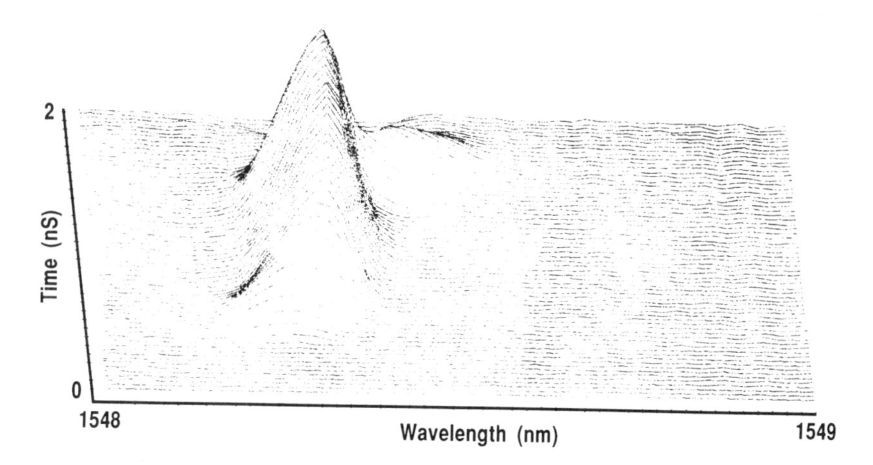

Figure 5.10. Experimental result showing a time dependent spectrum. The shift towards a longer wavelength is clearly seen.

By using a simple approximation we can derive an expression for the chirping from the rate equations. We assume that in the central part of the pulse the carrier density decreases linearly in time:

$$N = N_1 + \frac{dN}{dt} t . \qquad (5.39)$$

When this is combined with a linear dependence of the gain on the carrier density, the rate equation for the photon density, Eq. (5.8) (neglecting spontaneous emission and nonlinear gain), has a simple solution:

$$\frac{dS}{dt} = v_g \left(\Gamma a(N_1 - N_0) - (\alpha_{int} + \alpha_{end})\right) S + v_g \Gamma a \frac{dN}{dt} tS \,, \quad (5.40)$$

$$S = S_0 \exp\left(-\frac{(t-t_0)^2}{2\delta t^2}\right) . \quad (5.41)$$

This is simply a pulse that is Gaussian in time, and the width depends on the time derivative of the gain:

$$\delta t = \frac{1}{\sqrt{-v_g \Gamma a \frac{dN}{dt}}} = \frac{1}{\sqrt{-v_g \frac{dg_{net}}{dt}}} . \quad (5.42)$$

The full temporal width at half maximum is:

$$\Delta T = \sqrt{8 \ln 2}\, \delta t \approx \sqrt{\frac{8 \ln 2\, \Delta T}{-v_g \Delta g_{net}}} \,, \quad (5.43)$$

where Δg_{net} is the change in the gain during the time ΔT.

We recall the definition from Sec. 5.3 of the parameter α that relates changes in gain and changes in the refractive index:

$$\alpha = \frac{\Delta n_{eff}}{\Delta g_{net}} 2k \,, \quad (\alpha < 0) . \quad (5.44)$$

We can now relate the change in the gain to a change of the optical frequency [see also Eq. (5.28)]:

$$v_g \Delta g_{net} = -\frac{4\pi}{\alpha} \Delta f . \quad (5.45)$$

Combining this with Eq. (5.43) leads to a simple result for the time-bandwidth product:

$$|\Delta T\, \Delta f| = \frac{2 \ln 2}{\pi} |\alpha| . \quad (5.46)$$

The calculation can be carried out in a slightly more correct way by finding the spectral width from the Fourier transform of a linearly chirped Gaussian pulse:

$$|\Delta T\, \Delta f| = \frac{2 \ln 2}{\pi} \sqrt{1 + \alpha^2} . \quad (5.47)$$

The value of α is typically about -5, and we therefore find that the time bandwidth product for a semiconductor laser pulse is (at least) about five times higher than the transform limit. The limitation on the product of the transmission distance and the square of the data rate is consequently about five times more restrictive than the result based on the transform limit.

We assumed that the chirp was linear, but this is only true for the central part of the pulse. Consequently the spectrum of the pulse is not Gaussian, but it turns out that the expression for the width of the pulse that we derived agrees quite well with experimental results. We finally note that propagation of chirped pulses through dispersive fibers can give rise to some interesting effects.[22]

5.5. INTENSITY NOISE

Up to now we have considered the spontaneous emission rate to be a constant. However, the spontaneous emission is a random process, and we only know what the *average* rate is. This randomness gives rise to noise, and to study this phenomenon we look at the rate equations again. In the absence of a modulation current we have, from Eqs. (5.15) and (5.16):

$$j\,\omega\,\Delta N = -\frac{\Delta N}{\tau_s'} - \frac{\delta G'}{\delta N}\bar{S}\,\Delta N - G_{th}'\,\Delta S\,, \tag{5.48}$$

$$j\,\omega\,\Delta S = \frac{\delta G'}{\delta N}\bar{S}\,\Delta N - \gamma\,\Delta S + \frac{1}{V}F_s(\omega)\,. \tag{5.49}$$

The last term in Eq. (5.49) is the spectral density of the spontaneous emission noise. This term acts as the driving force (note that $F_s = 0$ leads to the trivial solution $\Delta N = \Delta S = 0$), and the laser is described as a noise driven oscillator. Strictly speaking there is a noise term in Eq. (5.48) as well, but it turns out that this can be neglected. Equations (5.48) and (5.49) are the rate equations in the frequency domain. Solving Eqs. (5.48) and (5.49), and using $H(\omega)$ from Eq. (5.20) and ω_r from Eq. (5.21) gives:

$$\Delta S(\omega) = H(\omega)\frac{\dfrac{1}{\tau_s'} + \dfrac{\delta G'}{\delta N}\bar{S} + j\omega}{\omega_r^2}\frac{1}{V}F_s(\omega)\,. \tag{5.50}$$

It is convenient to describe the noise in the form of a relative intensity noise. This quantity is defined as the power spectral density of the intensity noise at the frequency f, divided by the square of the average photon density. This is the relative noise one would measure per unit bandwidth by using a photodetector:

$$RIN = \frac{PDS(\Delta S)}{\bar{S}^2} . \tag{5.51}$$

In order to find the relative intensity noise we must first find the power spectral density of the spontaneous emission noise. Each spontaneous emission event adds a photon to the lasing field, but the corresponding phase is random. This is illustrated in Fig. 5.11, where we have introduced the number of photons in the laser S' as the product of the active volume V and the photon density S.

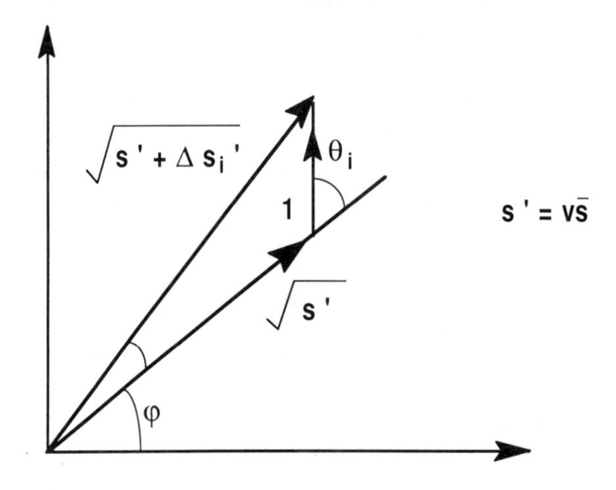

Figure 5.11. Change in the lasing field due to one spontaneous emission event.

Using the cosine relation

$$\sqrt{S'}^{\,2} + 1 + 2\sqrt{S'}\cos\Theta_i = \sqrt{S' + \Delta S_i'}^{\,2} , \tag{5.52}$$

we find the change in the photon number due to *one* spontaneous emission:

$$\Delta S_i' = 1 + 2\sqrt{S'}\cos\Theta_i . \tag{5.53}$$

The change in the photon density over a time t, due to spontaneous emission with a rate R, is found from Eq. (5.53). Since the average of $\cos\Theta_i$ gives 0:

$$\langle \Delta S' \rangle = \sum_{i=1}^{Rt} \Delta S_i' = Rt . \tag{5.54}$$

This agrees with the average spontaneous emission rate which appeared in the original rate equation (5.8):

$$\frac{d\langle\Delta S'\rangle}{dt} = R, \qquad \frac{dS}{dt}\Big|_{spont} = \frac{R}{V}. \tag{5.55}$$

Looking at the random part of Eq. (5.53), we now find that the total fluctuation after a time t, with spontaneous emission events occuring at the times t_i, is:

$$F_s(t) = \sum_{i=1}^{Rt} 2\sqrt{S'}\cos\Theta_i\,\delta(t-t_i). \tag{5.56}$$

We form the product

$$F_s(t)F_s(t') = 4S'\sum_{i=1}^{Rt}\cos^2\Theta_i\,\delta(t-t_i)\,\delta(t'-t_i), \tag{5.57}$$

and take the average of this expression:

$$\langle F_s(t)F_s(t')\rangle = 4S'\,\delta(t-t')\left\langle\sum_{i=1}^{Rt}\cos^2\Theta_i\,\delta(t'-t_i)\right\rangle. \tag{5.58}$$

The average of the square of the cosine gives 1/2, and we replace the sum by an integral (since there are a large number of spontaneous emissions). Hence

$$\langle F_s(t)F_s(t')\rangle = 2RS'\delta(t-t'). \tag{5.59}$$

We now assume that taking the ensemble average in Eq. (5.59) is equivalent with averaging over time (a process that satisfies this assumption is called ergodic). Equation (5.59) is then the autocorrelation function of $F_s(t)$. At this point we need the Wiener-Khintchine theorem, which states that the power spectral density of a function is the Fourier transform of its autocorrelation function. This gives, using the fact that the Fourier transform of the δ function is 1:

$$\mathrm{PSD}(F_s) = 2RS'. \tag{5.60}$$

We can now use Eqs. (5.50), (5.51), and (5.60) to find the relative intensity noise per unit bandwidth:

$$\mathrm{RIN} = |H(\omega)|^2\frac{1}{\omega_r^4}\left(\omega'^2+\omega^2\right)\frac{2R}{V\bar{S}}, \tag{5.61}$$

$$\omega' = \left(\frac{1}{\tau_s'} + \frac{\delta G'}{\delta N}\,\overline{S}\right). \qquad (5.62)$$

Since the resonance frequency ω_r is proportional to the square root of the photon density we have:

$$\mathrm{RIN} \propto \frac{1}{P_{out}^{3}} \qquad (5.63)$$

The relative intensity noise has a maximum near the resonance frequency:

$$\mathrm{RIN}(\omega_m=\omega_r) \propto \frac{1}{\gamma^2\overline{S}} \propto \frac{1}{\varepsilon^2 P_{out}^{3}}\ , \quad \text{for} \quad G_{th}'\,\varepsilon\,\overline{S} >> \frac{2R}{V\overline{S}}\ . \qquad (5.64)$$

This shows that a large value of the nonlinear gain parameter ε will reduce the intensity noise. The frequency dependence of the intensity noise is shown in Fig. 5.12.

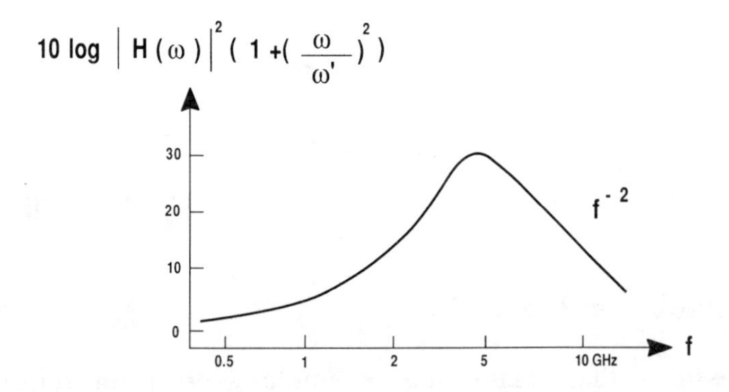

Figure 5.12 Frequency dependence of the relative intensity noise.

It should be noted that since the relative intensity noise per unit bandwidth depends on frequency it may vary significantly within a given bandwidth. In this case we must use the following expression for the total relative intensity noise between the frequencies f_1 and f_2:

$$\mathrm{RIN_T} = \frac{4R}{V\overline{S}} \int_{f_1}^{f_2} |\,H(\omega)\,|^2 \frac{1}{\omega_r^{4}}\left(\omega'^{\,2} + \omega^2\right) df\ . \qquad (5.65)$$

CHAPTER 6. LINEWIDTH

In order to derive an expression for the linewidth of a laser we first look at a simple representation of phase noise. Some differences between semiconductor lasers and other laser types are pointed out. Then, using the rate equations, the variance of the phase change is derived in a more rigorous way. This leads to a more precise explanation for the laser line shape. The last section of this chapter describes experimental methods for spectral measurements.

6.1. PHASE NOISE

In addition to the intensity noise described in Sec. 5.5, spontaneous emission also gives rise to phase noise. This eventually determines the linewidth of the lasing mode. A number of authors have published investigations of phase noise and linewidth for semiconductor lasers.[23-28] In this account we attempt to derive the important results as simply as possible.

Each spontaneous emission event leads to a phase change of the lasing field, as shown in Fig. 6.1.

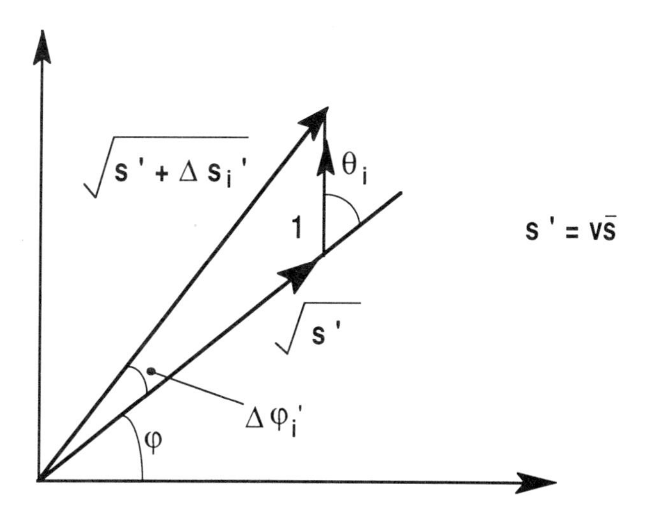

Figure 6.1. Change of magnitude and phase of the lasing field due to one spontaneous emission.

We remember from the previous section that the change of the photon number due to one spontaneous emission is given by:

$$\Delta S_i' = 1 + 2\sqrt{S'} \cos \Theta_i . \tag{6.1}$$

From Fig. 6.1 we also find the change of the phase of the optical field:

$$\Delta\phi_i' = \frac{1}{\sqrt{S'}} \sin \Theta_i \, . \tag{6.2}$$

After the photon number has been increased due to a spontaneous emission event it will relax back to the steady-state value. This will give a further phase change. The rate equation [Eq. (2.3), with R being the spontaneous emission rate] for the photon number is:

$$\frac{dS'}{dt} = (G' - \alpha') S' + R \, , \tag{6.3}$$

and the rate equation for the phase, Eq. (5.29), can be written:

$$\frac{d\phi}{dt} = -\frac{\alpha}{2}(G' - \alpha') + \text{constant} \tag{6.4}$$

$$= -\frac{\alpha}{2S'}\frac{dS'}{dt} + \text{constant} \, . \tag{6.5}$$

Note that α is the parameter introduced in Sec. 5.3, and α' is the loss per unit time.

A spontaneous emission event increases the photon number by the amount given in Eq. (6.1). The photon number must therefore subsequently decrease by the same amount in order to bring the laser back to steady state. Using Eqs. (6.4) and (6.5), this gives the following result for the additional phase change:

$$\Delta\phi_i'' = \frac{\alpha}{2S'}\Delta S_i' = \frac{\alpha}{2S'}(1 + 2\sqrt{S'} \cos \Theta_i) \, . \tag{6.6}$$

The total phase change is then:

$$\Delta\phi_i = \Delta\phi_i' + \Delta\phi_i'' \tag{6.7}$$

$$= \frac{\alpha}{2S'} + \frac{1}{\sqrt{S'}}(\sin \Theta_i + \alpha \cos \Theta_i) \, . \tag{6.8}$$

The average phase change over a time t due to a spontaneous emission rate R is (since the sine and cosine terms average to zero):

$$\langle\Delta\phi\rangle = \frac{\alpha}{2S'} Rt \, . \tag{6.9}$$

This corresponds to an angular frequency offset given by:

$$\Delta\omega = \frac{d\langle\Delta\phi\rangle}{dt} = \frac{\alpha R}{2S'}. \tag{6.10}$$

This is simply a constant frequency shift due to the average spontaneous emission rate.

Since the spontaneous emission process is random we have to look at the *variance* of the phase fluctuations. Using Eq. (6.8), noting that the first term is a constant, we find the variance after a time t:

$$\langle\Delta\phi^2\rangle = \frac{Rt}{S'}\left(\frac{1}{2} + \frac{1}{2}\alpha^2\right) = \frac{Rt}{2S'}(1 + \alpha^2). \tag{6.11}$$

We see that the variance increases *linearly* with time. Therefore the root-mean-square of the phase fluctuations $\sqrt{\langle\Delta\phi^2\rangle}$ increases with the *square root* of t. This result is expected since the phase fluctuation is a so-called random walk process. To illustrate this further, consider a random walk in the x-y plane. For each step in the positive x direction a step is also taken along the y axis, but it is random whether the step is in the positive or negative direction. Assume that the variance (defined as the *square* of the distance from the x axis) after N steps is Q:

$$\sigma_N^2 = Q. \tag{6.12}$$

We now want to find the variance after N+1 steps. There is a probability of 1/2 that the distance has increased by 1, and a probability of 1/2 that it has decreased by 1. Hence:

$$\sigma_{N+1}^2 = \frac{1}{2}(\sqrt{Q} + 1)^2 + \frac{1}{2}(\sqrt{Q} - 1)^2 = Q + 1. \tag{6.13}$$

Since the variance after one step is 1 we find:

$$\sigma_1^2 = 1 \quad \rightarrow \quad \sigma_N^2 = N. \tag{6.14}$$

This means that the deviation increases with $\sqrt{N}$. This behavior is illustrated in Fig. 6.2.

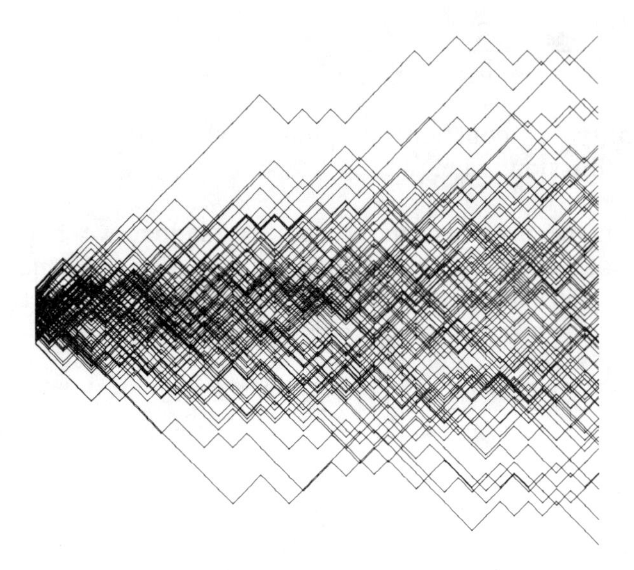

Figure 6.2. Computer simulated random walk, showing the increase in the deviation with the number of steps (40 steps for each curve, 100 curves).

6.2. LINEWIDTH AND LINE SHAPE

The lasing field is written as follows:

$$E(t) = \sqrt{S'}\,\exp(j\omega_0 t + j\phi(t)) \, , \tag{6.15}$$

where S' is the photon number, ω_0 is the center frequency, and ϕ is the phase. We form the autocorrelation function of the field, again assuming that we are dealing with an ergodic process (time averaging equivalent to ensemble averaging):

$$\langle E^*(t)\,E(t-\tau)\rangle = S' \left\langle \exp\big(j\phi(t-\tau) - j\phi(t)\big)\right\rangle \exp(j\omega_0\tau) \, . \tag{6.16}$$

This expression includes the phase change over the time τ, and we assume that the statistical properties of the phase change are time independent (i.e., the process is stationary):

$$j\phi(t-\tau) - j\phi(t) = j\Delta\phi(\tau) \, . \tag{6.17}$$

We know the variance of the phase change $\langle\Delta\phi^2\rangle$ from Eq. (6.11), and since the number of spontaneous emission events is large we can assume that the probability density function for the phase change is Gaussian. This follows from the central-limit theorem, which states that the probability distribution resulting from a convolution of a large number of identical probability distributions will approach a Gaussian distribution.

$$\mathrm{prob}(\Delta\phi) = \frac{1}{\sqrt{2\pi<\Delta\phi^2>}}\exp\left(\frac{-\Delta\phi^2}{2<\Delta\phi^2>}\right) . \tag{6.18}$$

Using this result in Eq. (6.16) gives (the proof is left as an exercise):

$$<E^*(t)\, E(t-\tau)> = S'\exp\left(-\frac{1}{2}<\Delta\phi^2>\right)\exp(j\omega_0\tau) . \tag{6.19}$$

Since we now have the autocorrelation function we can find the power spectral density for the laser mode by a Fourier transform. This gives the laser line shape:

$$L(\omega) = \frac{1}{1 + \dfrac{(\omega-\omega_0)^2}{\left(\dfrac{R}{4S'}(1 + \alpha^2)\right)^2}} . \tag{6.20}$$

The result is a Lorentz line shape with a full width at half maximum (the laser linewidth) given by:

$$\delta\nu = \frac{R}{4\pi S'}(1 + \alpha^2) . \tag{6.21}$$

The linewidth is proportional to the spontaneous emission rate and to the inverse of the photon number. Using the definition of R from Eq. (5.5), and the relation between the photon number S' and the output power per facet P_{out}, Eq. (2.6), we have, assuming that the reflectivities for the facets are the same:

$$\delta\nu = \frac{\hbar\omega\, v_g{}^2\, g_{th}\, \alpha_{end}\, n_{sp}}{8\,\pi\, P_{out}}(1 + \alpha^2) . \tag{6.22}$$

We note the inverse power dependence of the linewidth. The linewidth at an output power of 1 mW is typically in the range 50 to 100 MHz. This is in reasonable agreement with the result obtained by inserting the parameter values given in the symbol list into Eq. (6.22). Since α^2 can be quite large (a typical value of α is -5), this term is of major importance for the linewidth, and for obvious reasons it is generally known as the *linewidth enhancement factor*. A laser with a reduced value of this parameter will have a much reduced linewidth. In most other laser types $\alpha \approx 0$. The large value of n_{sp}, between 2 and 3, is also characteristic for semiconductor lasers.

A further correction factor to the linewidth is necessary. This correction is given by:[29-30]

$$K = K_t K_z , \tag{6.23}$$

$$K_t = \left| \frac{\iint |E(x,y)|^2 dx\,dy}{\iint E(x,y)^2 dx\,dy} \right|^2 , \quad K_z = \left| \frac{\int |E(z)|^2 dz}{\int E(z)^2 dz} \right|^2 . \tag{6.24}$$

The term K_t only differs from 1 when the laser structure does not have built-in lateral waveguiding. Such structures are sometimes referred to as "gain-guiding." The phase fronts in such structures are not planar, and a larger fraction of the spontaneous emission goes into the lasing mode. The field E(x,y) is complex and K_t can be much larger than 1. Most modern communication lasers have strong lateral guiding and therefore $K_t = 1$.

The factor K_z depends on the longitudinal field variation. In case of a low end loss (high facet reflectivity) the field is an almost perfect standing wave and K_z will be close to 1, but for low reflectivities the field differs from a pure standing wave. For a laser with field reflectivities r_1 and r_2 we have:

$$K_z = \left| \frac{(1 - r_1 r_2)\,(r_1 + r_2)}{2\, r_1 r_2 \ln\frac{1}{r_1 r_2}} \right|^2 \tag{6.25}$$

$$= \left(\frac{1 - R}{\sqrt{R}\, \ln\frac{1}{R}} \right)^2 \quad \text{for} \quad r_1 = r_2 = \sqrt{R}\,. \tag{6.26}$$

In the case of DFB lasers it is more complicated to find K_z. Since the position of the facet relative to the grating of a DFB laser cannot be controlled, the value of K_z can take a range of values because the field distribution depends on the facet position (Secs. 4.4 and 4.5). Some examples are shown in Fig. 6.3.

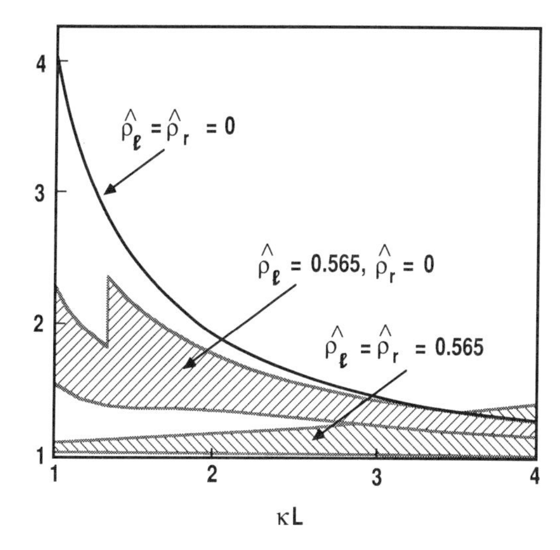

Figure 6.3. Linewidth correction factor K_Z for DFB lasers as function of the product of coupling coefficient (κ) and length (L). The facet field reflectivities are given by ρ_l and ρ_r (Ref. 31).

6.3. ALTERNATIVE DERIVATION OF LINE SHAPE

The result for the line shape can be derived in an alternative way that is more related to the derivation of intensity noise in Sec. 5.5. Including phase fluctuations in Eq. (5.29) we get an equation for the phase variation due to phase noise and variations in the carrier density:

$$\frac{d\phi}{dt} = F_\phi(t) - \frac{\alpha}{2}\frac{\delta G'}{\delta N}\Delta N(\omega) . \tag{6.27}$$

In Eq. (6.27) $\Delta N(\omega)$ is the spectral density of the fluctuations in the carrier density. The phase fluctuation due to spontaneous emissions in the time t is found from Eq. (6.2):

$$F_\phi(t) = \sum_{i=1}^{Rt} \frac{1}{\sqrt{S'}} \sin \Theta_i \, \delta(t-t_i) . \tag{6.28}$$

The phase change due to carrier density variations [the second part of Eq. (6.27)] can be found by substituting $\Delta N/\Delta S$ from Eq. (5.48) and ΔS from Eq. (5.50):

$$- \frac{\alpha}{2} \frac{\delta G'}{\delta N} \frac{\Delta N(\omega)}{\Delta S(\omega)} \Delta S(\omega) \tag{6.29}$$

$$= \frac{\alpha}{2} \frac{\delta G'}{\delta N} \frac{G_{th}'}{\frac{1}{\tau_s'} + \frac{\delta G'}{\delta N} \bar{S} + j\omega} H(\omega) \frac{\frac{1}{\tau_s'} + \frac{\delta G'}{\delta N} \bar{S} + j\omega}{\omega_r^2} \frac{1}{V} F_s(\omega) .$$

We can now find the power spectral density of the deviations of the optical frequency from its average value. For the PSD of the first term on the right-hand side of Eq. (6.27) we use a procedure similar to the treatment of photon number fluctuations [Eqs. (5.56) to (5.60)]. The second term, given by Eq. (6.29), can be significantly simplified. We then get :

$$\mathrm{PSD}\left(\frac{d\phi}{dt}\right) = \frac{R}{2S'} + \left(\frac{\alpha}{2S'}\right)^2 \mathrm{PSD}(F_s)\, |H(\omega)|^2 \cdot \tag{6.30}$$

$\mathrm{PSD}(F_s)$ is given by Eq. (5.60), and we finally have

$$\mathrm{PSD}\left(\frac{d\phi}{dt}\right) = \frac{R}{2S'}\left(1 + \alpha^2 |H(\omega)|^2\right) . \tag{6.31}$$

According to basic signal and noise theory[32] the variance of the phase is related to the power spectral density of the frequency fluctuation by:

$$\langle \Delta\phi^2 \rangle = \frac{t^2}{2\pi} \int_{-\infty}^{\infty} \mathrm{PSD}\left(\frac{d\phi}{dt}\right) \frac{\sin^2\left(\frac{\omega t}{2}\right)}{\left(\frac{\omega t}{2}\right)^2} d\omega . \tag{6.32}$$

It is obvious that the major contributions to the integral in Eq. (6.32) come from the low frequencies. If we simply use $|H(\omega)|^2 \approx |H(0)|^2 = 1$, Eq. (6.32) gives:

$$\langle \Delta\phi^2 \rangle \approx \frac{Rt}{2S'} (1 + \alpha^2) , \tag{6.33}$$

and the linewidth can be written:

$$\delta\nu \approx \frac{1}{2\pi}\mathrm{PSD}\left(\frac{d\phi}{dt}(\omega=0)\right) = \frac{R}{4\pi S'}(1 + \alpha^2)\,, \qquad (6.34)$$

which is the same as Eq. (6.21).

Because $|H(\omega)|$ has a peak near ω_r the true line shape [which is found from the Fourier transform of Eq. (6.19), after inserting Eq. (6.32)] will contain satellite peaks at $\omega = \omega_0 \pm n\omega_r$, where n = 1,2,... The reason for the difference between this and the result from section Sec. 6.2 is that in the derivation in Sec. 6.2 it was assumed that when the photon number increased it would relax back to the steady-state value immediately. In reality there is a delay, and the photon number will show damped oscillations (with frequency ω_r) before it settles down at the steady-state value. The satellite peaks in the line shape have been seen experimentally, as shown in Fig. 6.4.

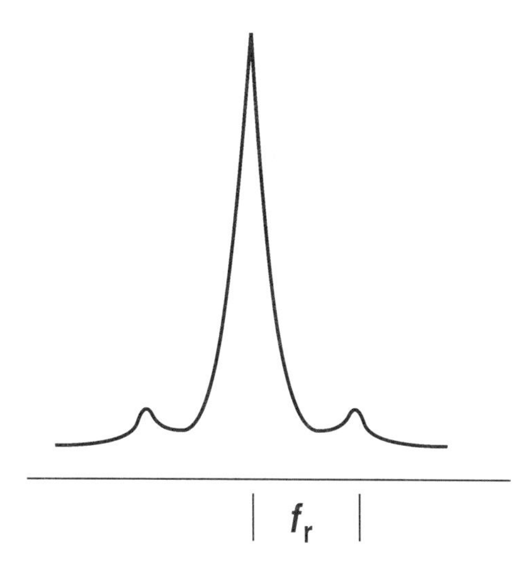

Figure 6.4. Experimental measurement of laser line shape, after Ref. 33.

6.4 SPECTRAL MEASUREMENTS

There are a number of methods that can be used to study the spectral properties of semiconductor lasers. Each method has its particular merits. The simplest form of spectral measurement involves the use of a spectrometer, shown schematically in Fig. 6.5.

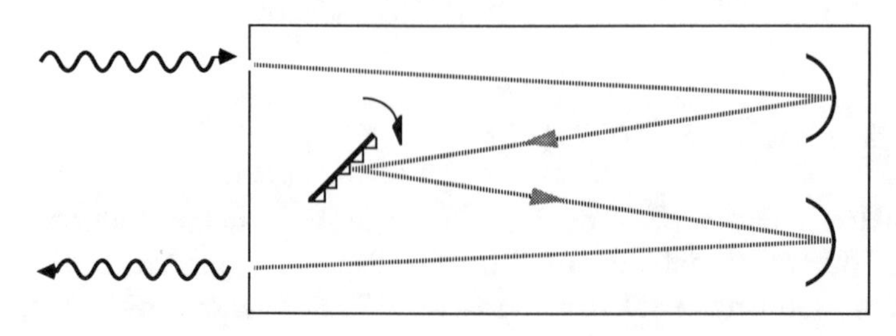

Figure 6.5. Spectrometer.

In the spectrometer the incoming light is dispersed by a grating, and only a narrow part of the spectrum passes through the exit slit. The spectrum can be traced by rotating the grating. The resolution of a spectrometer depends on the dispersive power of the grating, the length of the spectrometer, and the width of the exit slit. A particular advantage of a spectrometer is that the wavelength is measured in absolute terms. It is also well suited for multimode spectra. Since the resolution is typically about 0.1 nm (corresponding to about 12 GHz for wavelengths around 1.55 μm), a single semiconductor laser mode cannot be resolved. Figure 6.6 shows examples of laser spectra measured by a spectrometer.

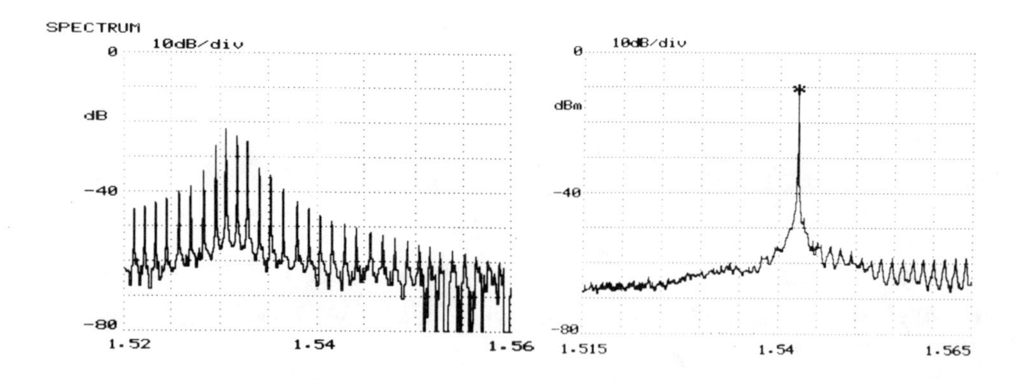

Figure 6.6. Spectrum of multimode laser (left) and DFB laser (right).

For a measurement with higher resolution, a Fabry-Pérot interferometer can be used. This instrument consists of two high reflectivity mirrors spaced a distance L. Interferometer resonances occur at frequencies spaced by c/2L (this spacing is called the free spectral range), and the transmission curve of the interferometer has peaks spaced by this frequency range, as shown schematically in Fig. 6.7.

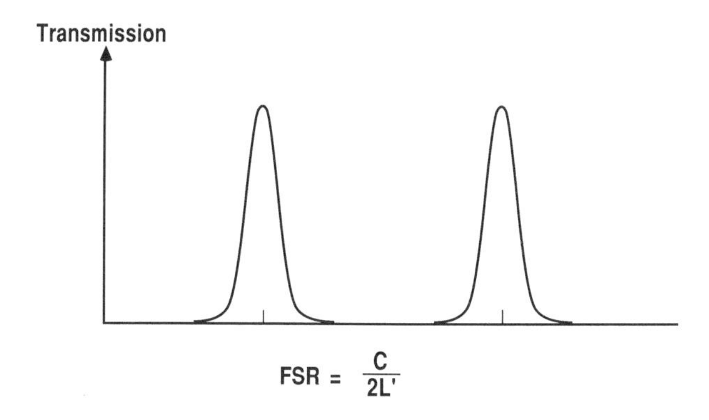

Figure 6.7. Transmission versus wavelength for a Fabry-Pérot interferometer.

By changing the mirror separation by using a piezoelectric translator the transmission peaks will scan through a range of frequencies (scanning Fabry-Pérot interferometer), and by detecting the transmitted light the optical spectrum can be displayed on an oscilloscope (see Fig. 6.8).

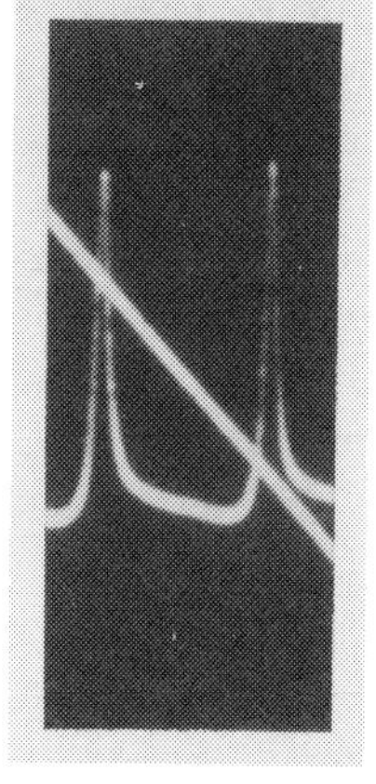

Figure 6.8. Spectrum measured by a Fabry-Pérot interferometer. The line across the figure is the driving voltage for the piezoelectric translator.

Because a Fabry-Pérot interferometer has multiple transmission peaks it is only useful for measurements of spectra narrower than the free spectral range. The resolution is typically in the 10 to 100 MHz range, depending on the length and mirror reflectivities. Though a semiconductor laser spectral line may not be completely resolved by a Fabry-Pérot interferometer, it is possible to study frequency modulation by looking at sidebands created by the modulation.

For high resolution measurements the so-called delayed self-heterodyne technique is particularly popular. The experimental setup for this method is outlined in Fig. 6.9.

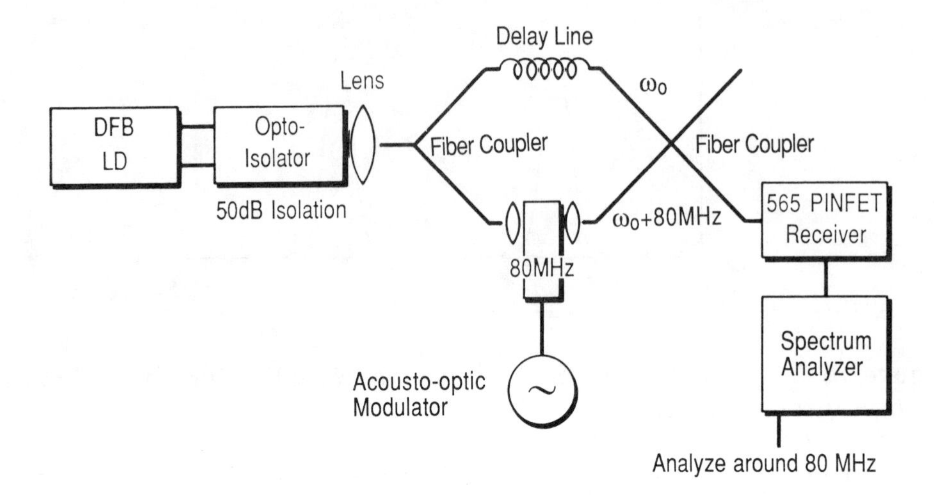

Figure 6.9. Delayed self-heterodyne measurement of linewidth.

The light from the laser is split in two parts. One part goes through a delay line, the other part is frequency shifted by an acousto-optic modulator. The two parts are recombined, the light is detected, and the detector signal is sent to a spectrum analyzer and displayed. If the delay is longer than the laser coherence time the spectral width of the detected signal will have the same spectral shape as the laser but it will be twice as wide. The delay caused by 1 km of optical fiber is about 5 μs, and if we require that the delay is longer than the coherence time (defined as 1 divided by the linewidth) we have:

$$\delta\nu > \frac{200 \text{ kHz km}}{L}, \tag{6.35}$$

where L is the length of the delay line. On the other hand the linewidth must be smaller than the frequency shift introduced by the modulator. Figure 6.10 shows some examples of measured spectra.

It turns out that this measurement technique can still be used even if the condition given by Eq. (6.35) is not fulfilled, but in this case the partial coherence of the two parts gives rise to a more complicated spectrum.[34]

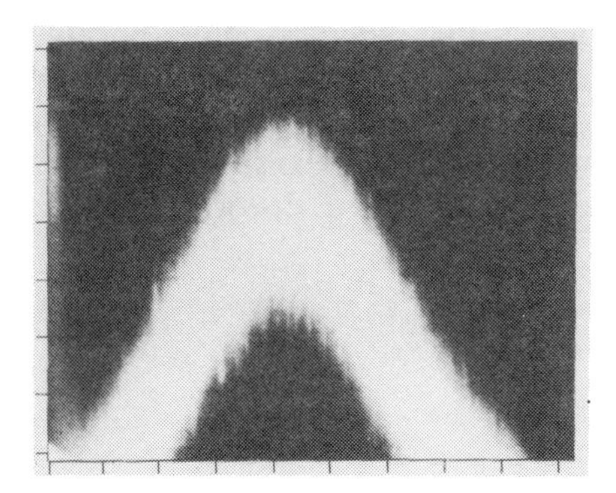

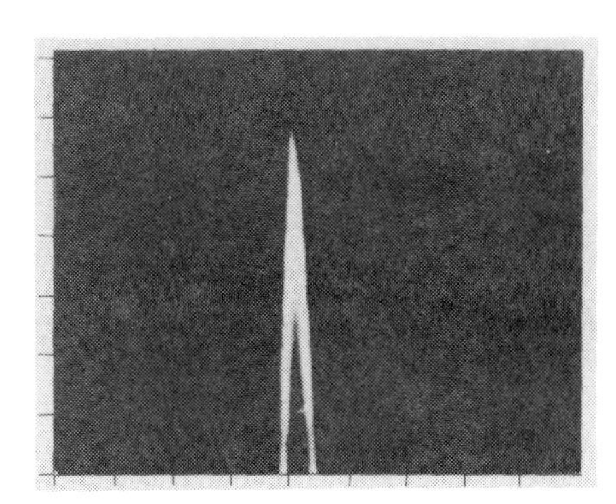

Figure 6.10. Spectral measurements by the delayed self-heterodyne technique. Solitary laser (left) and narrow linewidth external cavity laser (right). Vertical scale: 1 dB per division; horizontal scale: 10 MHz per division.

As previously mentioned, a semiconductor laser with an output power of 1 mW typically has a linewidth in the range 50 to 100 MHz. We also know that the linewidth decreases with increasing output power [Eq. (6.22)]. In Fig. 6.11 the linewidth is shown as function of $1/P_{out}$.

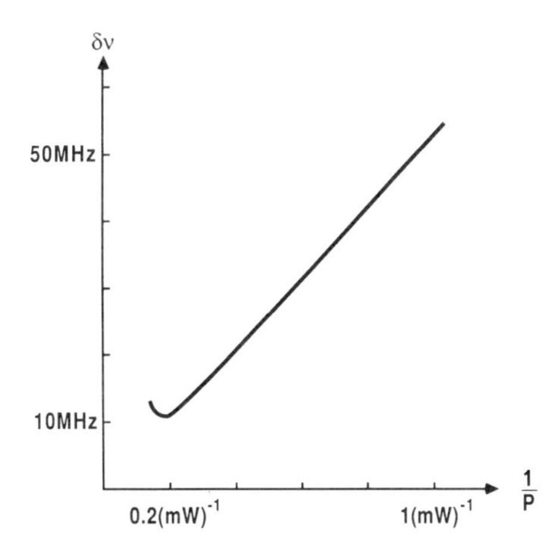

Figure 6.11. Linewidth as a function of the inverse output power for a DFB laser operating at room temperature at the wavelength 1.55 μm.

We can see from Fig. 6.11 that the measured linewidth has the expected value, and the expected power dependence up to an output power of about 5 mW. For higher power levels the linewidth first saturates and then rebroadens. The reason for this "linewidth floor" is not fully understood, but it is believed that spectral hole burning may be responsible. Spectral hole burning means that the stimulated recombination predominantly affects carriers in a certain energy range. If the carrier thermalization time is "slow," this spectral "hole" in the energy distribution of the carriers will give a distortion of the gain curve near the lasing wavelength.

CHAPTER 7. COHERENT SYSTEMS

The availability of single frequency semiconductor lasers makes coherent optical communication possible. In this chapter we first present the basic expressions for the signal-to-noise ratio. Then the relations between the signal-to-noise ratio and error probability are quoted for several digital modulation formats. Finally we discuss the laser requirements for coherent systems, with emphasis on laser linewidth.

7.1. SIGNAL-TO-NOISE RATIO

In a conventional direct detection optical communication system the optical signal field at the detector can be written:

$$E_S = \sqrt{P_S}\, \exp\big(j\omega_S t + j\phi(t)\big)\; m(t)\ . \tag{7.1}$$

Here P_S is the optical signal power, ω_S is the optical (angular) frequency, $\phi(t)$ is the optical phase, and $m(t)$ is the modulation function. In the case of digital modulation, $m(t)$ is 1 when a "1" is transmitted, and $m(t)$ is 0 when a "0" is transmitted. For $m(t) = 1$ the detector current can be written:

$$I_S = \frac{\eta e}{\hbar\omega} P_S\ , \tag{7.2}$$

where η is the detector efficiency, e is the unit charge, and $\hbar\omega$ is the photon energy. The expression in front of P_S in Eq. (7.2) is called the detector responsivity.

The receiver signal-to-noise ratio can be written:

$$\mathrm{SNR} = \frac{\left(\frac{\eta e}{\hbar\omega} P_S\right)^2}{2\,\frac{\eta e}{\hbar\omega}\, eBf_B\,(P_S + P_D) + \left(\frac{\eta e}{\hbar\omega} P_S\right)^2 \mathrm{RIN}_T}\ . \tag{7.3}$$

The numerator is the detected power, and the first term in the denominator is the shot noise due to the quantum nature of light, with B being the normalized bandwidth of the detector, and f_B the data rate. Noise sources due to the detector are described in terms of the equivalent detector noise power P_D. The last term is due to intensity noise of the signal, in addition to the shot noise (see Sec. 5.5).

In practice $P_D >> P_S$ and the SNR is limited by the detector noise. It is therefore not possible to achieve shot noise limited operation that requires $P_S >> P_D$.

Due to the statistical nature of the noise there is a finite probability that the noise is so large that it obscures the signal. The noise is usually assumed to have a Gaussian distribution, and in a system with an equal probability of "0" and "1" and the SNR defined by Eq. (7.3), the probability of an error (detecting a "1" when a "0" is transmitted and vice versa) is then given by the following expression:

$$P_e = \frac{1}{2}\,\mathrm{erfc}\left(\sqrt{\frac{SNR}{8}}\right), \tag{7.4}$$

where the complementary error function is defined by:

$$\mathrm{erfc}(x) = \frac{2}{\sqrt{\pi}}\int_x^{\infty} \exp(-t^2)dt \approx \frac{\exp(-x^2)}{\sqrt{\pi}\ x}, \text{ for large x.} \tag{7.5}$$

The standard requirement for a system is that the error probability is less than 10^{-9}. This is achieved for SNR $\approx$ 144, which for typical values of the receiver noise requires on the order of 2000 photons/bit.

In a coherent system the signal field $E_S = \sqrt{P_S}\exp\left(j\omega_s t+\phi(t)\right) m(t)$ is mixed with a local oscillator field $E_L = \sqrt{P_L}\exp(j\omega_L t)$, and for $m(t) = 1$ the detector current becomes:

$$2\frac{\eta e}{\hbar\omega}\sqrt{P_L P_S}\cos\left(\omega_{IF}t + \phi(t)\right), \tag{7.6}$$

where the intermediate frequency is:

$$\omega_{IF} = \omega_S - \omega_L. \tag{7.7}$$

The signal-to-noise ratio in this case is (neglecting the signal relative intensity noise):

$$SNR = \frac{2\left(\frac{\eta e}{\hbar\omega}\right)^2 P_S P_L}{2\frac{\eta e}{\hbar\omega} eBf_B\,(P_L + P_S + P_D) + \left(\frac{\eta e}{\hbar\omega}P_L\right)^2 RIN_T}. \tag{7.8}$$

For a large local oscillator power the noise terms due to the signal and the detector can be neglected, and if, in addition, the local oscillator intensity noise given by RIN_T is small we have:

$$SNR = \frac{\eta (Bf_B)^{-1}}{\hbar\omega} P_S . \qquad (7.9)$$

If the bandwidth is equal to the data rate (i.e., B = 1) then SNR = $\mathcal{N}$, where $\mathcal{N}$ is the number of *detected* photons per bit. Equation (7.9) is then the shot noise limit. The SNR is limited only by the quantum noise of the signal light.

7.2. MODULATION FORMATS

We have not yet specified the modulation function m(t) for coherent systems. There are a number of possibilities:

ASK: *A*mplitude *S*hift *K*eying:
m(t) = 1, for a "1"; m(t) = 0, for a "0"

FSK: *F*requency *S*hift *K*eying:
m(t) = $\exp(j\omega_m t)$, for a "1"; m(t) = 1, for a "0"

PSK: *P*hase *S*hift *K*eying:
m(t) = 1, for a "1"; m(t) = -1, for a "0"

The signal wave forms are shown in Fig. 7.1.

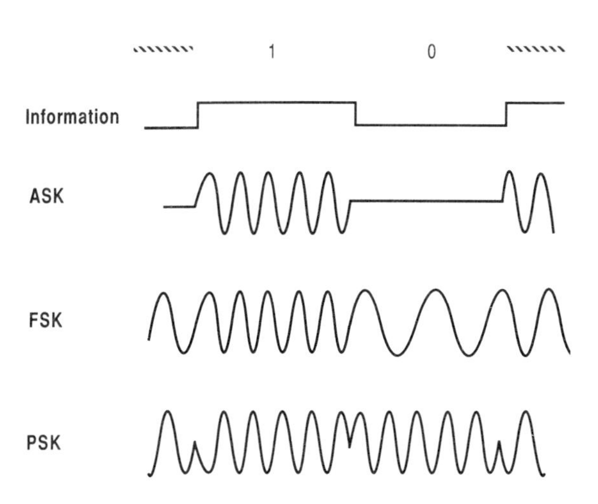

Figure 7.1. Wave forms for different modulation formats.

Two general classes of coherent systems exist: In *homodyne* systems the signal and local oscillator frequencies are identical, $\omega_L = \omega_S$, and the intermediate frequency is 0. This is achieved by locking the local oscillator frequency to that of the signal. In other words the optical signal is mixed directly down to baseband by the local oscillator. In contrast, in a *heterodyne* system the signal and local oscillator frequencies are different. Hence frequency locking is not required, but heterodyne systems require 3 dB more signal power to achieve the same error probability.

The analysis of optical coherent systems can be carried out exactly like the analysis of radio systems (see, for example, Ref. 35). In the case of shot noise limited operation and optimum receiver construction the error probabilities for the various systems are as shown in Table 7.1. The corresponding required numbers of detected photons are given in Table 7.2.

Table 7.1. Error probabilities for different (ideal) systems, given as functions of the number of detected photons per bit, N.

	Heterodyne	Homodyne
ASK	$\frac{1}{2}\,\mathrm{erfc}\left(\sqrt{\frac{N}{4}}\right)$	$\frac{1}{2}\,\mathrm{erfc}\left(\sqrt{\frac{N}{2}}\right)$
FSK	$\frac{1}{2}\,\mathrm{erfc}\left(\sqrt{\frac{N}{2}}\right)$	-
PSK	$\frac{1}{2}\,\mathrm{erfc}\left(\sqrt{N}\right)$	$\frac{1}{2}\,\mathrm{erfc}\left(\sqrt{2N}\right)$

Table 7.2 Required number of detected photons for a "1", for an error probability of 10^{-9}, for different (ideal) systems.

	Heterodyne	Homodyne
ASK	72	36
FSK	36	-
PSK	18	9

For ASK, power is only transmitted for "1" bits. Therefore, the average power for ASK is half the number given in Table 7.2 (provided that the probabilities for "1" and "0" are the same). The reduced signal power required for homodyne and PSK systems only comes at the expense of a more complicated receiver. Note that in all cases significantly less signal power is required than for a direct detection system.

The required number of photons for a "1" can be translated to a required signal power:

$$P_{req} = \left(-67 + 10 \log\left(\frac{\mathcal{N}}{\eta}\right) + 10 \log\left(\frac{f_B}{\text{Gbit/s}}\right) - 10 \log\left(\frac{\lambda}{\mu\text{m}}\right)\right) \text{dBm}. \tag{7.10}$$

We must bear in mind that the number of photons given above is the theoretical minimum, which assumes ideal wave forms and a high value for the local oscillator power. In practice a higher number is needed. As an example of required power we consider a near ideal FSK system with $\mathcal{N} = 45$, $\eta = 0.8$, $f_B = 140$ Mbit/s, and $\lambda = 1.55$ µm. In this case the required signal power is -60 dBm (=1 nW).

In addition to the modulation formats mentioned above there are other possibilities, such as

DPSK: *D*ifferential *P*hase *S*hift *K*eying:
the phase is only changed if one symbol differs from the previous one.

CPFSK: *C*ontinuous *P*hase *F*requency *S*hift *K*eying:
this is like FSK, but the phase is continuous from one symbol to the next; this is not necessarily the case for FSK.

MSK: *M*inimum Frequency *S*hift *K*eying:
can be regarded as the limiting case of CPFSK; the frequency difference between a "1" and a "0" corresponds to one half optical cycle over one bit period.

Coherent systems offer two major advantages over direct detection systems. First, as we have just seen, the required signal power is smaller. This means that a signal of a given power can be transmitted over a longer distance. Second, the receiver is only sensitive to signals in a narrow frequency range, so only optical signals satisfying $\omega_S - \omega_L \approx \omega_{IF}$ will be detected. This means that the very large optical bandwidth offered by an optical fiber can be used much more efficiently, since it is possible to transmit a large number of optical frequencies over the same fiber, provided these frequencies are spaced by a few times ω_{IF}.

7.3. LASER REQUIREMENTS

From the preceding discussion we can see that there are a number of requirements which a laser must fulfill before it can be used in a coherent system:

1. The frequency of the local oscillator laser, ω_L, must be adjustable in order to keep the intermediate frequency ω_{IF} fixed. This can be done by temperature or current tuning.

2. A high local oscillator power is required ($P_L >> P_D$). This is particularly important at high data rates since the equivalent detector noise P_D increases with increasing frequency.

3. The intensity noise (RIN) of the local oscillator laser should be low. Notice, however, that the intensity noise according to the results in Sec. 5.5 depends on the laser power.

4. For ASK and PSK systems external modulation may be necessary, but for FSK and CPFSK systems the signal source can be a directly modulated laser. Frequency modulation of lasers is described in Sec. 5.3. The main practical problem is the nonuniform FM response (see Figs. 5.6 and 5.7). One way of overcoming this problem is to use an equalization circuit to attenuate the low frequency part of the signal spectrum.

5. The linewidths of both the signal laser and the local oscillator laser have to be narrow. The linewidth requirement is illustrated in more detail for two specific cases in the following.

In Fig. 7.2 we look at a block diagram for a "dual filter/dual detector" FSK system.

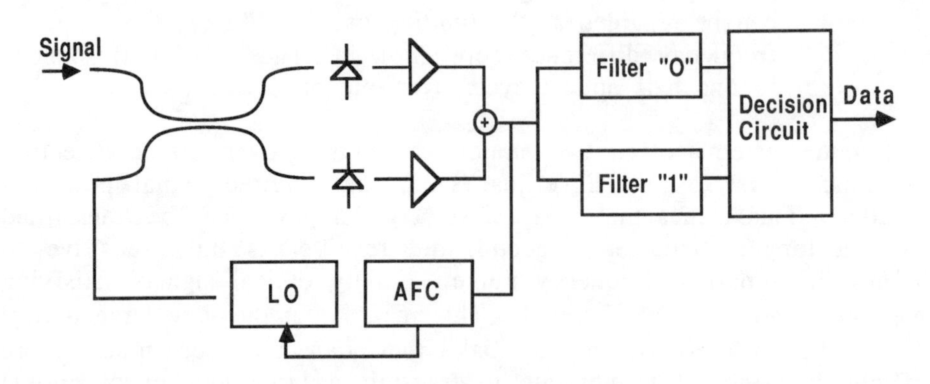

Figure 7.2. Block diagram for FSK system.

The signal and local oscillator fields are combined in a fiber coupler. In principle a single detector would be sufficient, but in this case a part of either the signal or the local oscillator power would be lost. A further advantage of the dual detector configuration is that the intensity noise from the local oscillator laser can be at least partly cancelled.[36] The local oscillator frequency tracks the signal frequency by means of an automatic frequency control circuit.

In an FSK system the symbol "0" corresponds to one frequency and the symbol "1" corresponds to a different frequency. The decision whether a detected signal is a "0" or a "1" is made by comparing the output from the two filters. Alternatively, the output from a single filter can be compared to a reference level. This is also the method used in ASK systems.

In order to study the influence of laser phase noise (i.e., laser linewidth) we first look at the transfer functions for the filters in the receiver (Fig. 7.3).

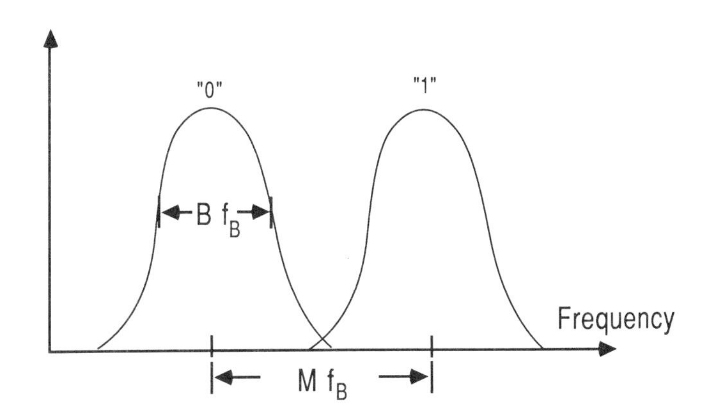

Figure 7.3. Magnitude of the transfer functions for the filters in an FSK system.

The filter for "0" is centered around $\omega_{IF}/2\pi$ and has a bandwidth normalized to the data rate of B. The filter for "1" is centered around $(\omega_{IF} + \omega_m)/2\pi$, $\omega_m = 2\pi$ M f_B , where M is the modulation index.

We now describe the phase noise in terms of a frequency offset. If the optical phase changes by $\Delta\phi$ during the time T, this frequency offset is simply defined as the phase change averaged over T:

$$\delta f = \frac{\Delta\phi}{2\pi T}. \tag{7.11}$$

If the frequency offset caused by the phase noise is so large that the signal appears in the frequency range for the wrong filter, a detection error will occur. Setting the time T in Eq. (7.11) equal to the inverse of the filter width (i.e., $1/Bf_B$) the error probability becomes:

$$P_e = \text{prob}\left(\delta f > \frac{M}{2} f_B\right) = \text{prob}\left(\Delta\phi > \frac{\pi M}{B}\right). \tag{7.12}$$

Using the probability density function for $\Delta\phi$ from Eq. (6.18) and the variance of the phase deviation during the time T, $<\Delta\phi^2> = 2\pi\ \delta\nu\ T$, where $\delta\nu$ is the linewidth, the error probability is:

$$P_e = \frac{1}{2}\,\text{erfc}\left(\frac{\frac{\pi M}{B}}{\sqrt{4\pi\delta\nu\frac{1}{Bf_B}}}\right) = \frac{1}{2}\,\text{erfc}\left(\sqrt{\frac{\pi M^2 f_B}{4\delta\nu B}}\right). \tag{7.13}$$

Requiring the error probability to be less than 10^{-9} we have:

$$\frac{\pi\ M^2\ f_B}{4\ \delta\nu\ B} > 18\,, \tag{7.14}$$

which leads to the following linewidth requirement:

$$\delta\nu < \frac{\pi\ M^2}{72\ B} f_B \quad (= 0.17\ f_B\,,\quad \text{for } M = 2,\ B = 1)\,. \tag{7.15}$$

It is characteristic that the allowed linewidth is related to the data rate. For a higher data rate, lasers with a larger linewidth can be used. It is seen from Eq. (7.15) that a larger value of M (the modulation index) allows a larger linewidth, but use of a larger M means that the "1" filter is at a higher frequency and the detector must be able to handle higher frequencies. The fact that the receiver noise power P_D increases with frequency means that a higher local oscillator power is necessary in order to achieve shot noise limited performance.

For the single filter case an error occurs if the phase noise takes the signal outside the filter range. In the case of a sharp filter with a bandwidth Bf_B this leads to:

$$\delta\nu < \frac{\pi\ B}{72} f_B\,. \tag{7.16}$$

A larger linewidth can therefore be used if the filter is widened, but from Eq. (7.8) we see that this leads to a decrease in the signal to noise ratio. It turns out that by appropriate signal processing filter widening with only a small deterioration in the SNR is possible.[37]

For a DPSK system an error occurs if the phase fluctuation over one bit period is more than $\pi/2$:

$$P_e = \mathrm{prob}\left(\Delta\phi > \frac{\pi}{2}\right). \tag{7.17}$$

The appropriate time T is the inverse of the data rate and we find:

$$P_e = \frac{1}{2}\,\mathrm{erfc}\left(\frac{\frac{\pi}{2}}{\sqrt{4\,\pi\,\delta\nu\,f_B^{-1}}}\right). \tag{7.18}$$

An error probability less than 10^{-9} is obtained for

$$\frac{\pi\,f_B}{16\,\delta\nu} > 18\,, \tag{7.19}$$

which gives a very stringent linewidth requirement:

$$\delta\nu < \frac{\pi}{288}\,f_B = 0.011\,f_B\,. \tag{7.20}$$

It must be remembered that in these linewidth estimates we have only considered the errors caused by phase noise (finite linewidth). These errors will occur even in the limit of infinite signal power, and these limiting cases are therefore referred to as the "error rate floor." Errors also occur if the signal power is too low compared with the total noise, and in a more complete analysis both effects must be considered (see Fig. 7.4).

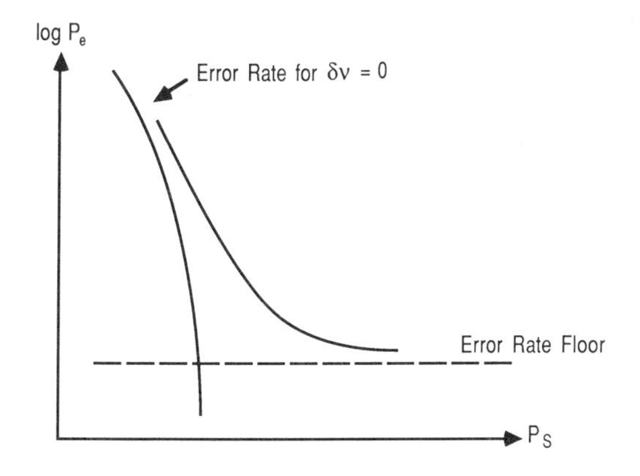

Figure 7.4. Error probability as a function of signal power. The figure also shows the limiting case of zero linewidth and the error rate floor caused by a finite linewidth.

A finite linewidth causes a "power penalty", i.e., an increase in the signal power necessary to achieve a certain error probability. In practice it is therefore necessary to have the "error rate floor" due to the finite linewidth well below 10^{-9}, and the linewidth requirements are therefore somewhat more stringent than the results we have derived. See for example Ref. 38 for a detailed analysis of ASK and FSK systems.

In the equations in this section $\delta\nu$ is the *combined* linewidth of signal and local oscillator (i.e., the IF linewidth). If the lasers have the same linewidth, the linewidth of each of them must be half the results derived.

Some typical examples of practical linewidth requirements are given in Tables 7.3 and 7.4.

Table 7.3. Maximum linewidth relative to the data rate, $\delta\nu/f_B$ (assuming the signal and local oscillator laser to have identical linewidths).

System		IF linewidth	Laser linewidth (identical lasers)
I	DPSK	0.003	0.0015
II	FSK, narrow deviation	0.07	0.035
III	FSK, wide deviation	0.4	0.2

Table 7.4. Maximum laser linewidth in MHz for typical data rates (assuming the signal and local oscillator laser to have identical linewidths).

System		Data rate 140 Mbit/s	560 Mbit/s	2 Gbit/s
I	DPSK	0.2	0.8	3.0
II	FSK, narrow deviation	4.9	19.6	70.0
III	FSK, wide deviation	28.0	112.0	400.0

A standard DFB laser can satisfy the linewidth requirements for a number of these system/data rate combinations, but for DPSK systems and for low data rates, lasers with narrower linewidths must be used.

CHAPTER 8. NARROW LINEWIDTH LASERS

As seen in Chap. 7 the application of semiconductor lasers in coherent systems can place stringent requirements on the linewidth. In this chapter we discuss methods for reducing the laser linewidth. Use of external cavities is an interesting possibility, but the behavior of such devices is rather complicated, and in some cases undesirable phenomena occur. Finally some examples of external cavity lasers are described.

8.1. FACTORS AFFECTING THE LINEWIDTH

It is clear from the preceding section on coherent systems that for some cases standard DFB lasers with a linewidth in the 10 to 100 MHz range cannot be used. We will therefore investigate how the laser linewidth can be reduced. We first recall the formula for the laser linewidth derived in Sec. 6.2:

$$\delta\nu = \frac{\hbar\omega\, v_g{}^2\, g_{th}\, \alpha_{end}\, n_{sp}}{8\,\pi\, P_{out}}(1+\alpha^2)\,K\,. \qquad (8.1)$$

The threshold gain is given by:

$$g_{th} = \alpha_{int} + \alpha_{end}\,, \qquad (8.2)$$

where α_{int} is the internal loss and α_{end} is the end loss. For a Fabry-Pérot laser the facet loss is:

$$\alpha_{end} = \frac{1}{L}\ln\left(\frac{1}{R}\right), \qquad (8.3)$$

and for a DFB laser with a high κL (κ is the coupling coefficient) it can be approximated by Eq. (4.26):

$$\alpha_{end}L = 2\alpha_o L \approx 2\left(\frac{\pi}{\kappa L}\right)^2. \qquad (8.4)$$

It is therefore clear that a longer laser will have a narrower linewidth. This can also be seen directly from Eq. (6.21). In a longer laser the photon number S' will be higher and the lower facet loss will give a lower value of the spontaneous emission rate R [Eq. (5.5)]. This argument, however, is not quite correct, since we have considered the linewidth for a given output power P_{out}. But when the facet loss is decreased the efficiency, which is given by

$$\eta = \frac{1}{2}\frac{\alpha_{end}}{\alpha_{int} + \alpha_{end}}, \tag{8.5}$$

will also decrease. The laser will therefore have to be pumped harder in order to maintain the same output power.

Another possibility for reducing the linewidth is to reduce the value of α. A possible reduction of this parameter will have a significant influence on the linewidth, which depends on α^2. Looking at Fig. 5.5 we can see that there is a photon energy dependence in the changes in gain and refractive index. Therefore the value of α will depend on the photon energy. This is shown in more detail in Fig. 8.1.

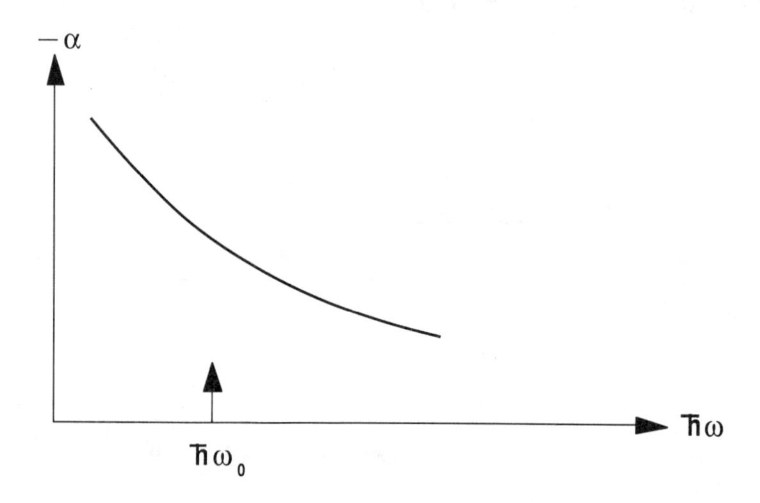

Figure 8.1. Variation of α with photon energy. The maximum gain occurs at $\hbar\omega_0$.

A conventional laser will always operate at the maximum gain, but a DFB laser can be forced to operate at a wavelength different from that corresponding to the gain peak. This is done by deliberately "detuning" the Bragg wavelength of the grating away from the gain peak. From Fig. 8.1 we see that α decreases with increasing photon energy (decreasing wavelength). A narrower linewidth is therefore obtained by forcing the laser to operate at a shorter wavelength ($\lambda_{Bragg} < \lambda_{gmax}$). This is called "negative detuning." An example is shown in Fig. 8.2.

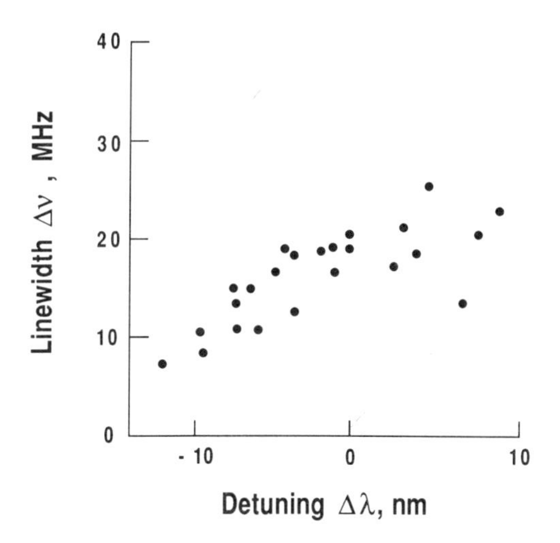

Figure 8.2. Linewidth in MHz as a function of detuning in nm. The detuning is defined as $\lambda_{Bragg} - \lambda_{gmax}$ (Ref. 39).

A different way to reduce α is to use a quantum well structure. In such a structure the active region consists of one or more very thin (≤ 100 Å) active layers. In such thin layers the carriers cannot move freely in the direction perpendicular to the layer structure, but their momentum in this direction can only take on certain values. This situation is known from the quantum mechanical particle-in-a-box problem, and explains the origin of the name quantum well. Due to this quantum confinement in one dimension the density of states function is changed, and this again leads to a different gain curve. The result is that the values of both n_{sp} and $|\alpha|$ are reduced, leading to a narrower linewidth. An example is shown in Fig. 8.3.

By combining laser length, detuning, and the use of quantum wells several groups have reported linewidths under 1 MHz. Some of the best results have been obtained using long, phase-shifted multi-quantum well structures: 270 kHz[41] and 250 kHz[42]. In a quantum well laser with a varying grating pitch, a linewidth of 170 kHz[43] has been obtained. These results are probably close to the limit of what it is possible to achieve in a relatively simple solitary laser structure.

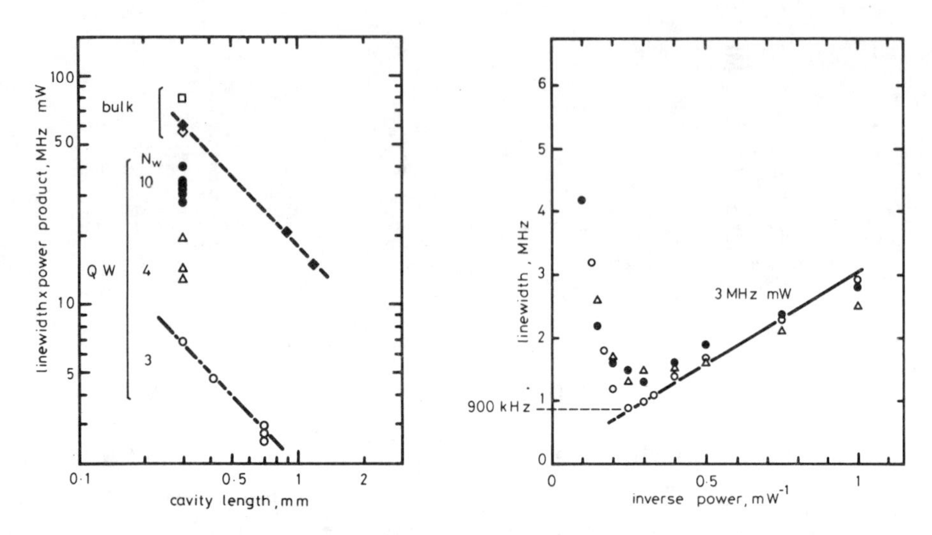

Figure 8.3. Linewidth results for quantum well DFB lasers working at a wavelength of 1.5 μm. The left-hand side of the figure shows the linewidth power product as a function of cavity length. Nw is the number of quantum wells in the structure. The right-hand side shows the linewidth as a function of inverse power for three 700 μm long quantum well lasers (Ref. 40).

8.2. EXTERNAL CAVITIES

Substantial linewidth narrowing can be achieved by using external cavities. We have already looked at coupled cavities in Sec. 3.3; the initial stages of the analysis of the linewidth of external cavity lasers follow the same lines. A mirror with a (field) reflectivity r_e is placed to the right of the right-hand laser facet as shown in Fig. 8.4.

The combined reflectivity due to the facet and the external mirror is:

$$r_2'' = r_2 + \frac{(1-r_2^2)\, r_e \exp(-j\omega\tau_e)}{1 + r_2\, r_e \exp(-j\omega\tau_e)}, \tag{8.6}$$

where τ_e is round-trip time in the external cavity. If multiple reflections are neglected this expression can be simplified to:

$$r_2'' \approx r_2\left(1 + \frac{1-r_2^2}{r_2} r_e \exp(-j\omega\tau_e)\right). \tag{8.7}$$

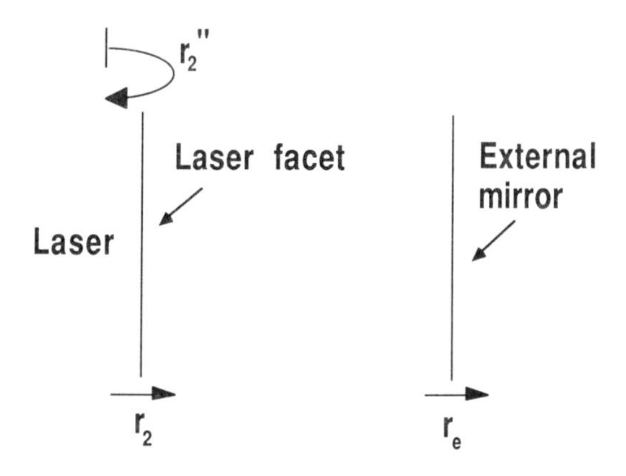

Figure 8.4. External cavity. A mirror is placed outside the laser cavity.

The total loss is now:

$$v_g\left(\alpha_{int} + \frac{1}{L}\ln\left(\frac{1}{r_1 r_2''}\right)\right), \tag{8.8}$$

where L is the laser length. If the second term in Eq. (8.7) is small the loss can be written:

$$G_{th}' - \frac{v_g}{L}\frac{1-r_2^2}{r_2} r_e \exp(-j\omega\tau_e) = G_{th}' - \frac{2}{\tau_d}\frac{1-r_2^2}{r_2} r_e \exp(-j\omega\tau_e)\,, \tag{8.9}$$

where τ_d is the round-trip time in the laser. The second term represents the change in threshold due to the external cavity. We are now in a position to write a rate equation for the *field*. Spontaneous emission is neglected, and it is assumed that the laser, in the absence of the external cavity, will operate at the (angular) frequency ω_o, with the field given by $E_o \exp(j\omega_o t)$. Including the external feedback we have:

$$\frac{dE}{dt} = \left(j\omega_o + \frac{1}{2}\left(G'(N) - G_{th}'\right)(1 - j\alpha)\right)E(t) + \frac{1}{\tau_d}\frac{1-r_2^2}{r_2} r_e\, E(t) \exp(-j\omega\tau_e). \tag{8.10}$$

The factor 1/2 on the gain is included because we are dealing with the *field* rather than with the intensity. The α parameter appears because changes in the carrier density (which are related to changes in the gain G') will lead to changes in the optical frequency.

We now assume that the laser will operate at an angular frequency ω (different from ω_o):

$$E(t) = E_o \exp(j\omega t) \cdot \qquad (8.11)$$

When this is inserted into Eq. (8.10) we get:

$$j\omega = j\omega_o + \frac{1}{2}\Delta G' (1 - j\alpha) + \frac{1}{\tau_d}\frac{1-r_2^2}{r_2} r_e \exp(-j\omega\tau_e) . \qquad (8.12)$$

The real and imaginary parts of this expression are, respectively:

$$\Delta G' = -2\frac{1}{\tau_d}\frac{1-r_2^2}{r_2} r_e \cos(\omega\tau_e) , \qquad (8.13)$$

$$\omega - \omega_o = -\frac{\alpha}{2}\Delta G' - \frac{1}{\tau_d}\frac{1-r_2^2}{r_2} r_e \sin(\omega\tau_e) . \qquad (8.14)$$

These can be combined to give the frequency change due to the presence of the external cavity:

$$\Delta\omega = \omega - \omega_o = \frac{1}{\tau_d}\frac{1-r_2^2}{r_2} r_e \left(\alpha \cos(\omega\tau_e) - \sin(\omega\tau_e)\right) . \qquad (8.15)$$

This can also be written:

$$\Delta\omega\, \tau_e = -\frac{\tau_e}{\tau_d}\frac{1-r_2^2}{r_2} r_e \sqrt{1 + \alpha^2}\, \sin\left(\omega\tau_e - \arctan(\alpha)\right) . \qquad (8.16)$$

The influence of the external cavity can be described in terms of the parameter C:

$$C = \frac{\tau_e}{\tau_d}\frac{1-r_2^2}{r_2} r_e \sqrt{1 + \alpha^2} . \qquad (8.17)$$

Taking the derivative of Eq. (8.16) with respect to ω_o gives

$$\frac{d\omega}{d\omega_o} = 1 - C \cos\left(\omega\tau_e - \arctan(\alpha)\right)\frac{d\omega}{d\omega_o} , \qquad (8.18)$$

$$\frac{d\omega}{d\omega_o} = \frac{1}{1 + C\cos\left(\omega\tau_e - \arctan(\alpha)\right)} . \qquad (8.19)$$

This means that fluctuations in the lasing frequency ω due to fluctuations in the "internal" frequency ω_o will be damped by the factor given by Eq. (8.19). Since the linewidth depends on the square of such fluctuations we get the following relation between the linewidth of the external cavity laser $\delta\nu_e$ and the linewidth of the solitary laser $\delta\nu_o$:

$$\delta\nu_e = \delta\nu_o\left(\frac{d\omega}{d\omega_o}\right)^2 . \qquad (8.20)$$

We now assume that the laser will operate at a frequency where the threshold gain has its minimum value. This means that the phase of the reflected field will have a value that makes $-\Delta G'$ in Eq. (8.13) as large as possible ($\omega\tau_e = 2p\pi$, where p is an integer). In this case we derive the following from Eqs. (8.17), (8.19), and (8.20):

$$\delta\nu_e = \delta\nu_o\left(\frac{1}{1 + \frac{\tau_e}{\tau_d}\frac{1-r_2^2}{r_2}r_e}\right)^2 . \qquad (8.21)$$

It should be noted that the minimum linewidth due to feedback will occur for a different value of the phase of the reflected field [this phase is determined from $\omega\tau_e = \arctan(\alpha)$]. In this simple analysis we have not considered the stability properties.

We note from Eq. (8.21) that the linewidth $\delta\nu_e$ depends on the *square* of the external cavity length, which is proportional to τ_e. This is what we expect since the linewidth for a laser with a small internal loss also depends on the square of the length [see Eq. (8.1)].

8.3. FEEDBACK REGIMES

The analysis in Sec. 8.2 was carried out assuming a relatively low level of feedback (multiple reflections were neglected). In general the behavior of a laser with an external cavity is quite complicated and depends on both the magnitude of the reflection and the length of the external cavity. Stability and noise properties of external cavity lasers have not yet been completely investigated.

External cavity behavior has been classified into five regimes:[44]

I. If the parameter C, defined in Eq. (8.17), is smaller than 1 the external cavity laser will show stable operation, but the linewidth will depend on the phase of the reflected field.

II. For C > 1 there are multiple solutions for $\omega\tau_e$ from Eq. (8.15). This can lead to jumping between two external cavity modes with closely spaced wavelength. In this case the time-averaged spectrum of the laser will have two peaks, each corresponding to one of the possible modes.

III. As C increases further a regime is reached where the laser operates in a stable single mode with a narrow linewidth. This happens because the mode where the feedback phase gives the narrowest linewidth becomes more stable than the mode corresponding to the lowest threshold[45] [note that this differs from the situation leading to Eq. (8.21)].

IV. Above a certain level of feedback the linewidth suddenly broadens by several orders of magnitude. This phenomenon was called "coherence collapse" in Ref. 46. A laser operating in this regime is unsuitable as a narrow linewidth source. The coherence collapse can occur when C exceeds a critical value which is given by $0.5\ \gamma\,\tau_e$, where γ is the damping factor [see Eqs. (5.17) and (5.22)] and τ_e is the round-trip time in the external cavity.[47] In this regime there is also a strong increase in the intensity noise of the laser.

V. For very strong feedback, stable single mode operation with a very narrow linewidth occurs again. In this case the properties of the external cavity system is determined more by the external cavity than by the laser.

In Fig. 8.5 we show an example of theoretical results for the linewidth properties of an external cavity laser. In these calculations noise due to spontaneous emission has been simulated numerically.

In general the boundaries between the different feedback regimes will depend on the operating conditions and on a number of laser parameters. Important parameters are external cavity length, laser power, laser length, and the value of the α parameter.

An example of experimental results are shown in Fig. 8.6.

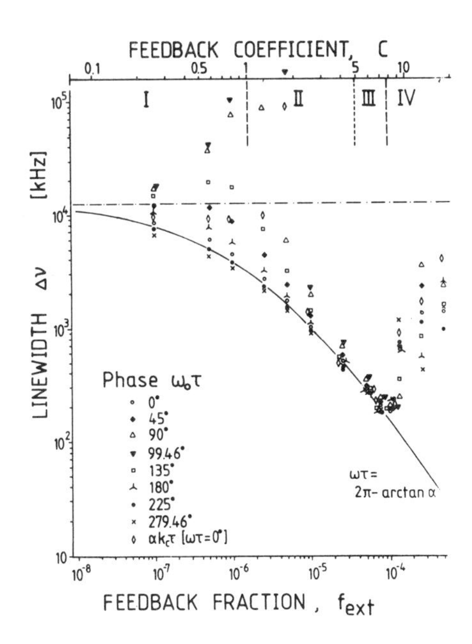

Figure 8.5. Linewidth as a function of the external power reflectivity $R_e = r_e^2$. In this example the laser power is 5 mW, the external cavity is a fiber of length 10 cm (index 1.5), and $\alpha = 6$. The different points correspond to different feedback phases, and the solid line is the minimum linewidth. The feedback regimes are indicated (Ref. 48).

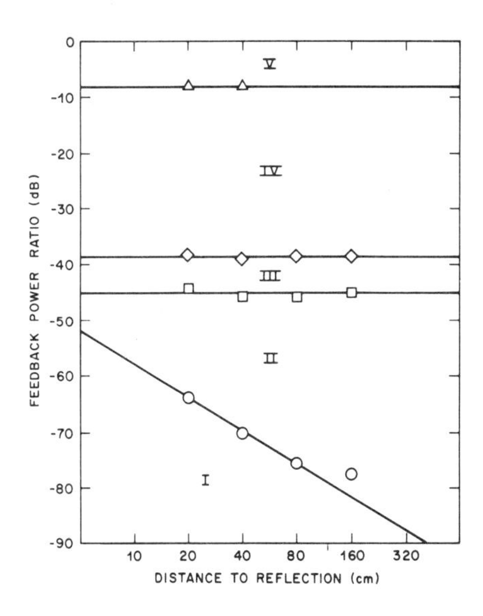

Figure 8.6. Boundaries between feedback regimes for a DFB laser (Ref. 44).

8.4. EXAMPLES OF EXTERNAL CAVITY LASERS

Long external cavities may lead to very narrow linewidths. The external reflector can either be a mirror or a grating. In the latter case it is possible to tune the laser mechanically simply by rotating the grating. In order to suppress internal laser modes the laser facet facing the external cavity should be antireflection coated.

In Fig. 8.7 we show an alternative arrangement where the external cavity is an optical fiber. A wavelength selective reflection is achieved by etching a grating into the fiber. Due to the high reflectivity of the grating a linewidth of 3 kHz was obtained.

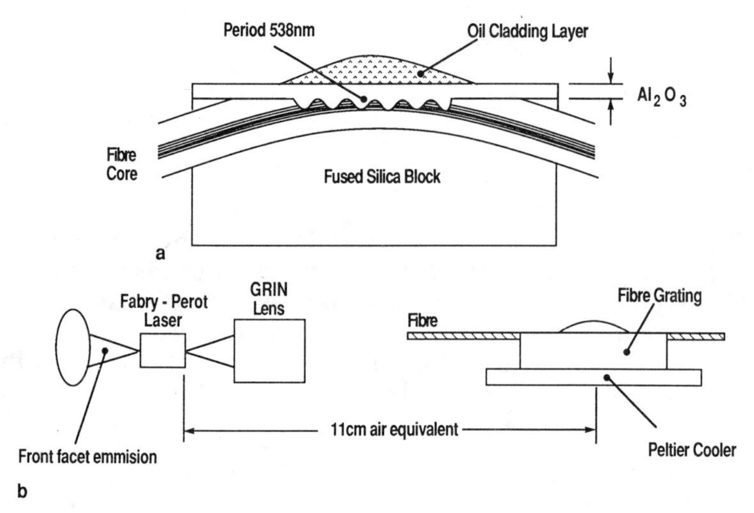

Figure 8.7. Fiber-grating external cavity laser (Ref. 49).

The very narrow linewidth possible with long external cavities comes at the expense of decreased separation between the external cavity modes, and creates potential stability problems. In addition it is not possible to frequency modulate lasers with long external cavities if the data rate is similar to, or exceeds, the external cavity mode spacing since this would lead to mode hopping.

As we saw in Sec. 7.3 even a moderate level of linewidth narrowing may be of interest for coherent systems. Relatively short external cavities may therefore be useful. They will also allow frequency modulation.

An example of a short external cavity is the graded index rod lens shown in Fig. 3.2. In Fig. 8.8 experimental linewidth results are shown for such a structure.

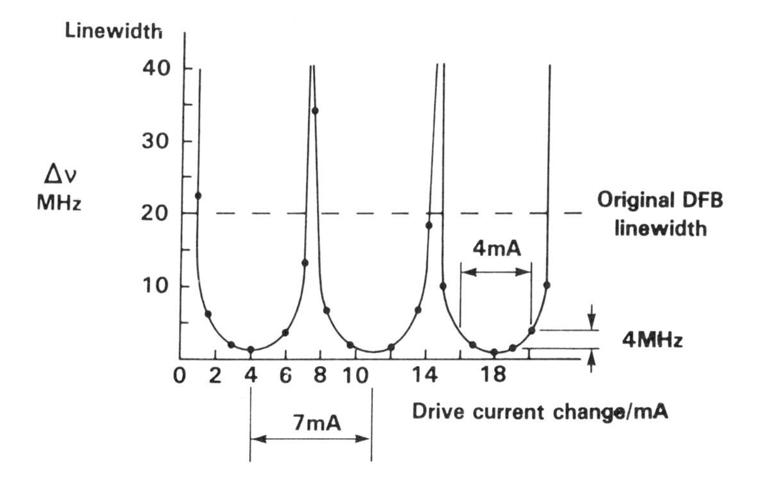

Figure 8.8. Linewidth as a function of current change of a DFB laser coupled to a graded index rod lens external cavity.

The periodic variation of the linewidth seen in Fig. 8.8 is due to a change of the phase of the reflected signal because the optical frequency is current dependent. In this case the optical frequency changes by about 1.5 GHz/mA, and the period of 7 mA (≈ 10 GHz) corresponds to the external cavity round-trip length (≈ 3 cm).

An even more compact device can be fabricated by integrating the external cavity with the laser, as shown in Fig. 8.9.

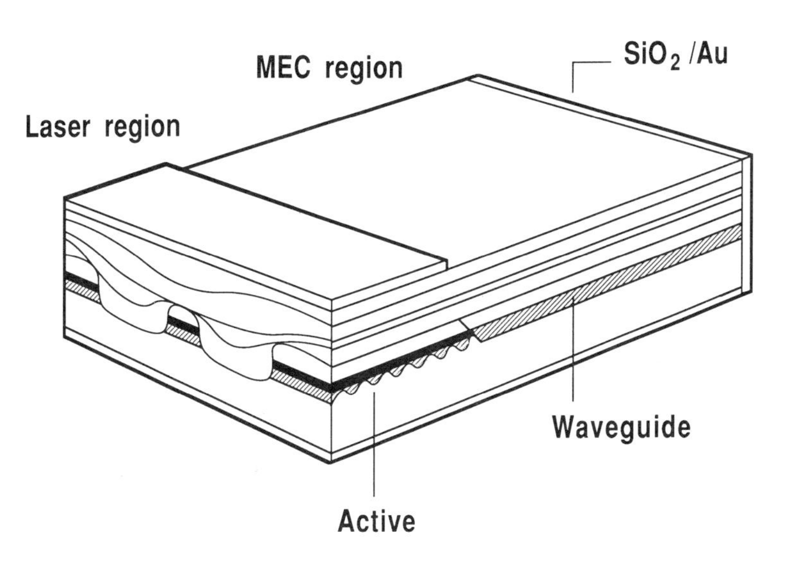

Figure 8.9. Integrated passive cavity (IPC) laser (Ref. 50).

It follows from the analysis of external cavities that the linewidth reduction G, which can be obtained in an IPC laser, is given by:

$$G \approx \eta_c^2 \left(\frac{L_e}{L}\right)^2 r_e^2 \exp(-2\ \alpha_e\ L_e)\ . \tag{8.22}$$

In this expression η_c is the coupling efficiency between the active region and the waveguide, L_e is the length of the passive section, L is the length of the active section, r_e is the reflectivity at the end of the passive section, and α_e is the loss in the passive section. Since $\eta_c \leq 1$ and $r_e \leq 1$ the maximum linewidth reduction is:

$$G \leq \left(\frac{L_e}{L}\right)^2 \exp\left(-2\ (\alpha_e L)\ \frac{L_e}{L}\right)\ . \tag{8.23}$$

This relation is illustrated in Fig. 8.10.

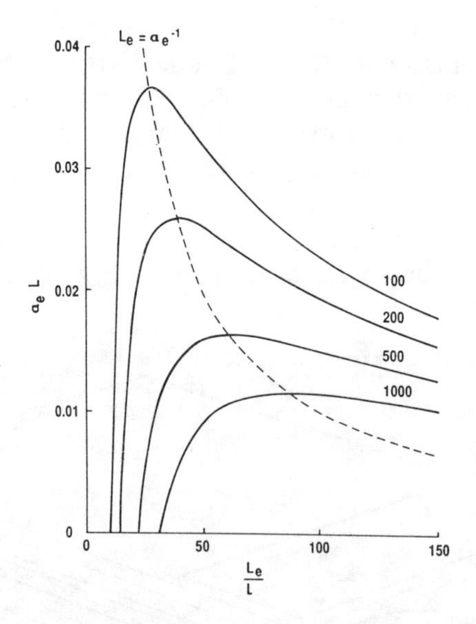

Figure 8.10. The curves give values for the maximum linewidth reduction in an IPC structure as a function of $\alpha_e L$ and L_e/L (Ref. 51).

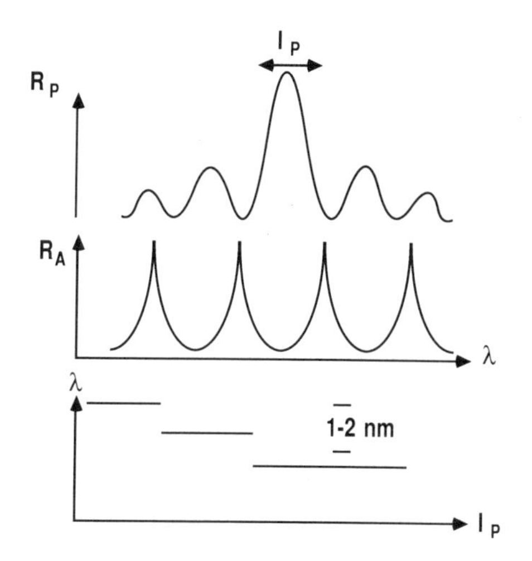

Figure 9.4. Tuning of multisection laser with a passive DBR section.

In Fig. 9.2 the tuning was continuous, but only over a short range. In Fig. 9.4 a wider tuning range was achieved, but there were gaps in the tuning range. In order to obtain both a wide tuning range and avoid gaps, a three-section structure, which combines the tuning characteristics shown in Figs. 9.2 and 9.4, can be used (see Fig. 9.5).

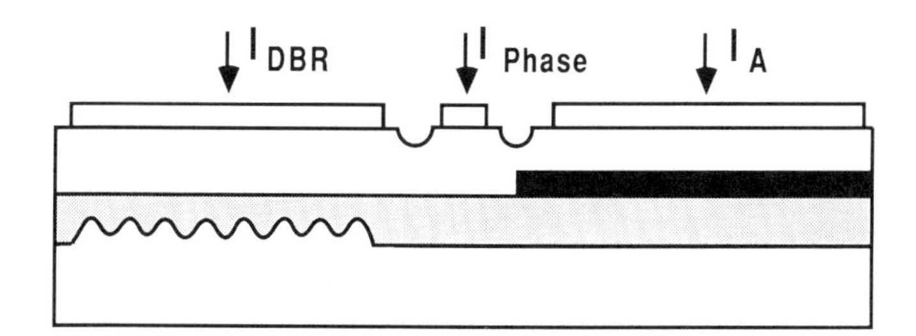

Figure 9.5. Three-section structure with separate DBR and phase control sections.

In this structure all wavelengths can be accessed by using a combination of two tuning currents, as shown in Fig. 9.6. There is now a wide wavelength coverage with no gaps, but two control currents are required, and the tuning is not continuous over the whole range.

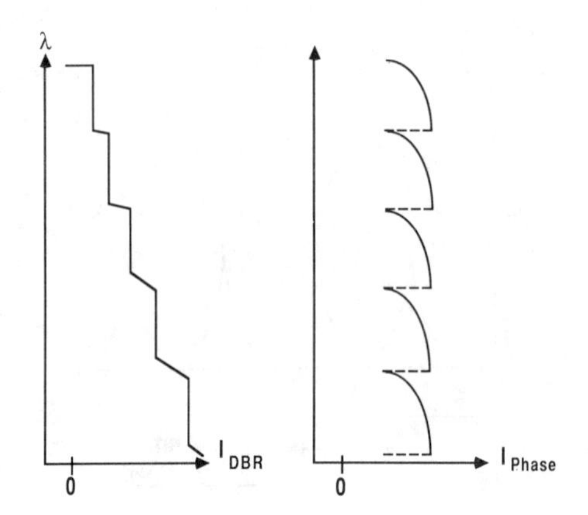

Figure 9.6. Tuning of a three-section structure. The total tuning range is 5.8 nm (Ref. 52).

A simpler control of the wavelength is obtained if *one* control current is divided between the DBR and phase control sections. This works as follows: When the current to the DBR section is increased the refractive index in this region decreases, and the DBR reflection will occur at a shorter wavelength. The phase shift for one round trip through the rest of the structure is $4\pi(L_{act}n_{act} + L_{phase}n_{phase})/\lambda$. This phase shift increases with decreasing wavelength, and to compensate for this n_{phase} is reduced by an increase in the current to the phase control section. When the balance between the DBR current and the phase control current is correct the round-trip phase change remains constant and the wavelength is tuned continuously over several nm by one control current.

As an alternative to multisection lasers it has been proposed to have the tuning region and the active layer in parallel, by placing a tuning layer above the active layer as shown in Fig. 9.7.

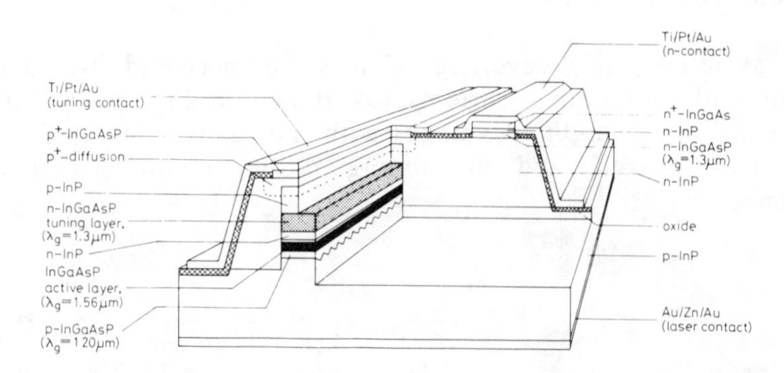

Figure 9.7. Tunable twin-guide (TTG) DFB laser (Ref. 53).

Results from TTG lasers have shown a continuous tuning range of over 7 nm.[54]

9.2. FABRICATION OF MULTISECTION LASERS

Fabrication of multisection lasers such as the ones described in the previous section requires multiple growths and rather complicated processing. An example of the fabrication procedure for a multisection laser is illustrated in Figs. 9.8(a)-(i) which show the fabrication step-by-step.

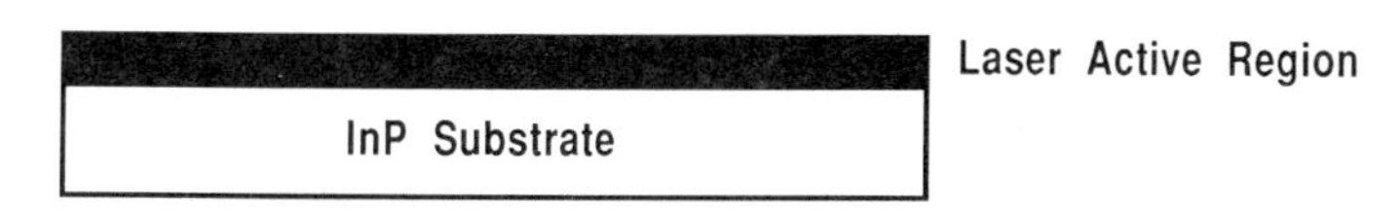

Figure 9.8(a). Grow source wafer.

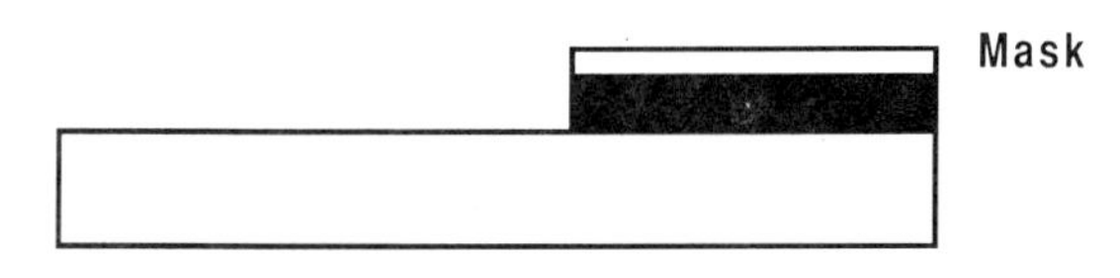

Figure 9.8(b). Deposit and define mask; etch out tuning section.

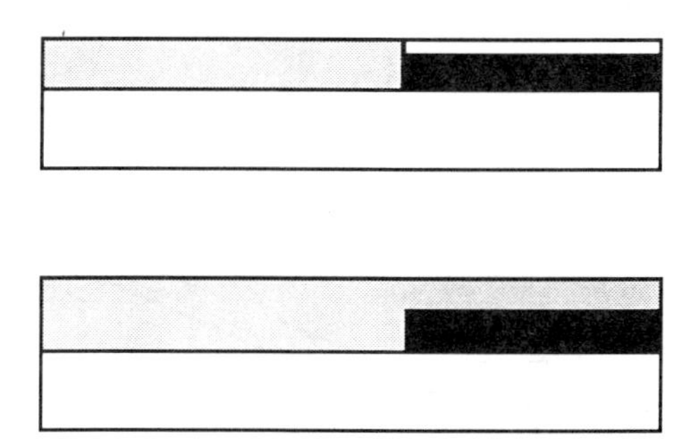

Figure 9.8(c). Infill tuning section, selectively (top) or nonselectively (bottom).

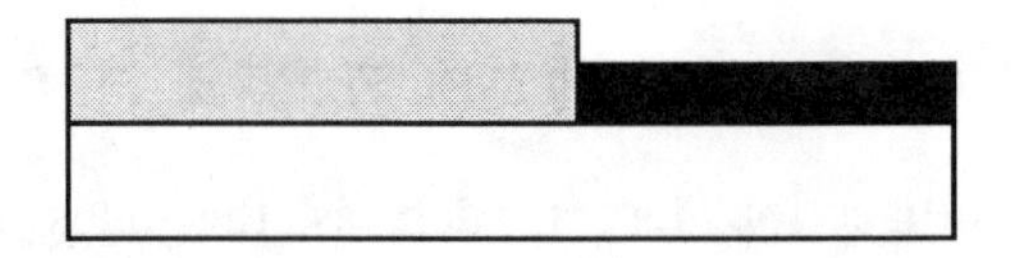

Figure 9.8(d). Mask off and etch off excess infill over the active region (for nonselective infilling). Alternatively carry out a diffusion step over the active region. This step is necessary since the material for the tuning section is undoped (for low loss), while good conductivity is required over the active region.

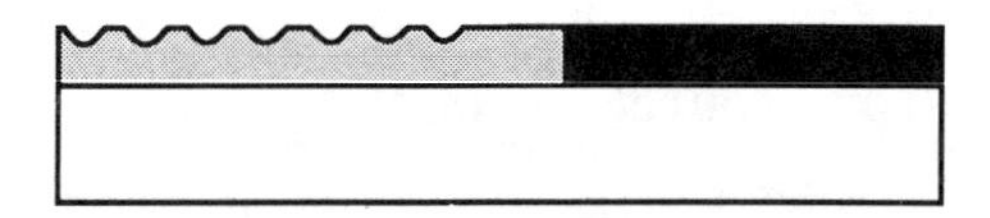

Figure 9.8(e). Define the grating in the Bragg reflector section.

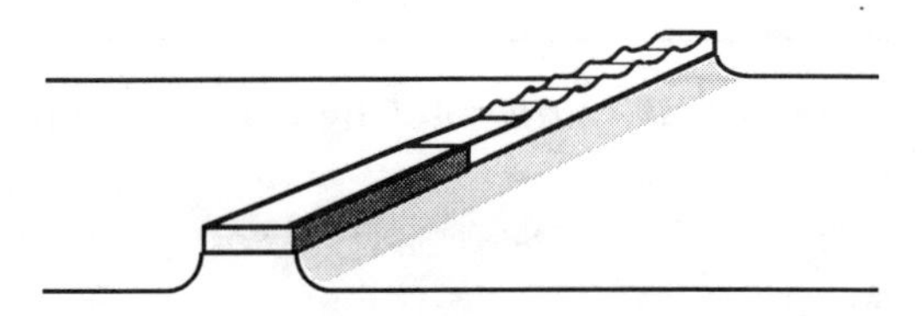

Figure 9.8(f). Etch laser stripe.

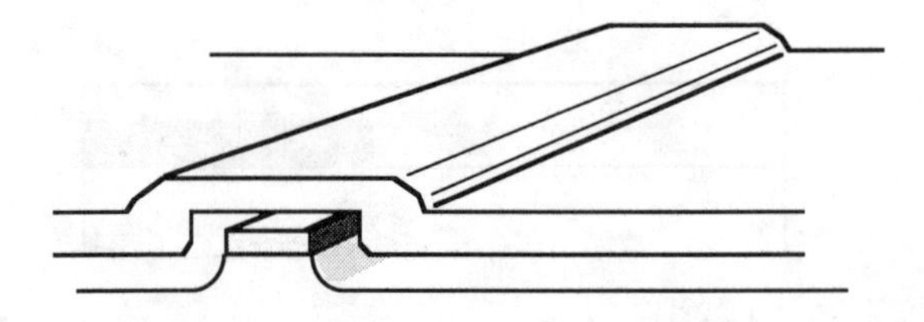

Figure 9.8(g). Overgrow current blocking layers.

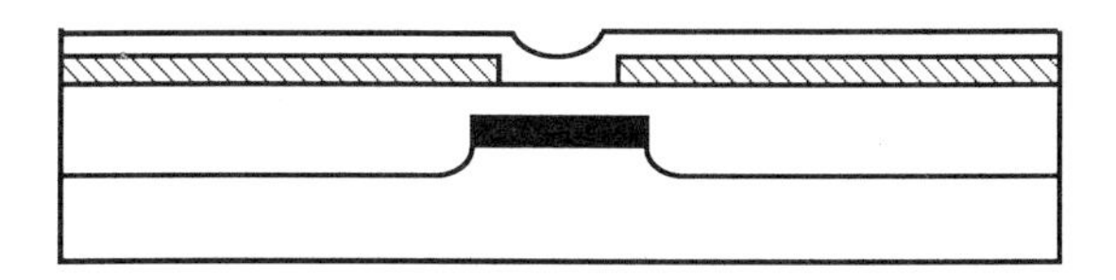

Figure 9.8(h). Define dielectric passivation and top contact metallization.

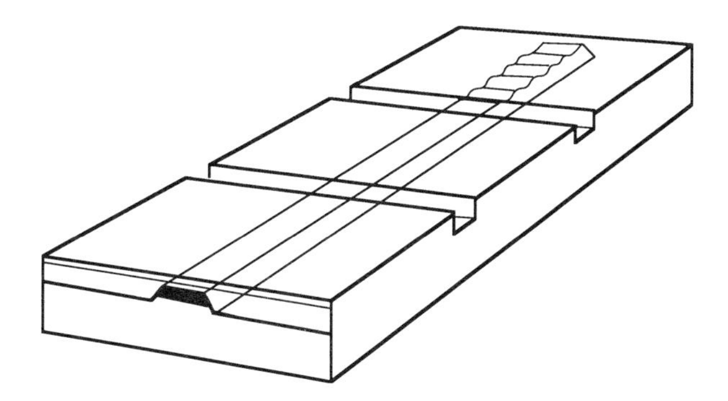

Figure 9.8(i). Define isolation slots.

The remaining processing steps are similar to those for ordinary laser structures.

9.3. OPTOELECTRONIC INTEGRATED CIRCUITS (OEICs)

Multisection lasers can be regarded as a step towards more complicated devices where several optoelectronic functions are integrated in the same structure. Below we show two examples of such devices. The device in Fig. 9.9 contains three separate lasers and three Bragg reflectors. The Bragg reflectors are fabricated in such a way that the reflection peaks occur at slightly different wavelengths. This is achieved by having diffent layer thicknesses in the Bragg sections. The outputs from the lasers are combined in a passive waveguide. In order to compensate for the losses in the waveguide and in the joints an amplifier (which can be regarded as a laser operating below its threshold) is placed at the end of the device. The device is thus able to transmit information carried on different optical wavelengths (wavelength division multiplexing, or WDM).

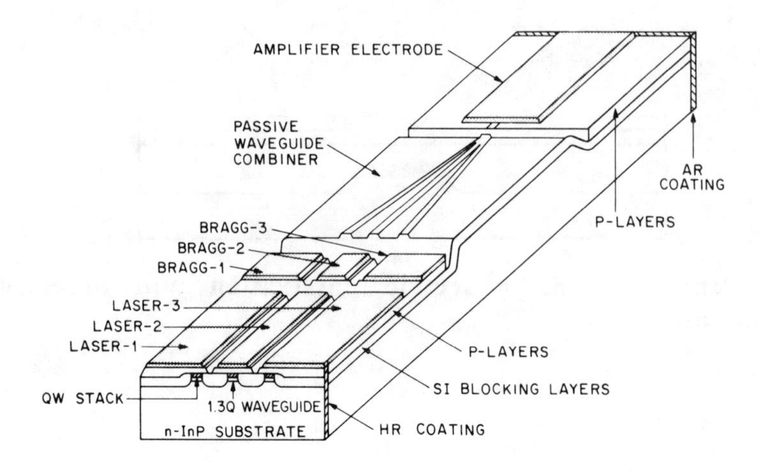

Figure 9.9. OEIC for WDM transmitter (Ref. 55).

In Fig. 9.10 we show an integrated receiver for a coherent system. This device contains a tunable local oscillator laser, an input waveguide, a coupler, and a dual detector.

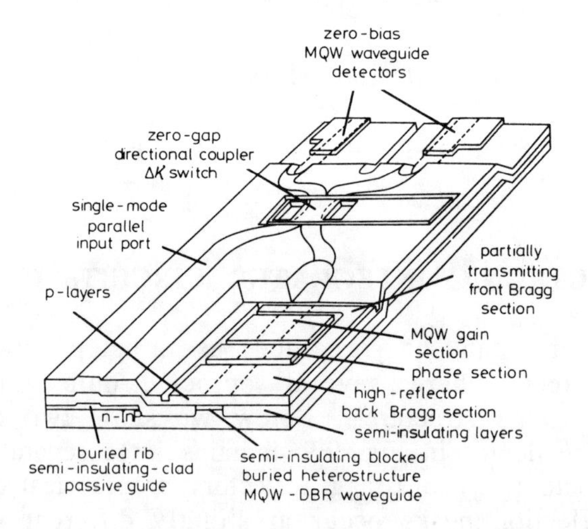

Figure 9.10. Coherent receiver OEIC (Ref. 56).

CHAPTER 10. CONCLUSION AND OUTLOOK

One of the major areas of application for semiconductor lasers is optical fiber communication. The emergence of lasers with higher and higher performance has made increasingly sophisticated communication systems possible, and there is a constant demand from the systems engineers for even higher laser performance.

Most optical communication systems are still using multimode lasers, but single frequency lasers will reduce the dispersion and thus make transmission at higher speed and over longer distances possible. With the emergence of single frequency lasers with a linewidth of under 100 MHz it was realized that coherent optical communication might be possible, with the potential for higher receiver sensitivity and better utilization of the extremely high bandwidth offered by optical fibers. This in turn has led to the requirement for widely tunable lasers for use in multichannel systems-the "optical radio."

In addition to narrow linewidth and tunability other requirements include high power, high modulation speed, and good reliability.

There is little doubt that improvement of laser performance will continue for a number of years, and that new demands will arise as the current problems are solved. Some of the issues that will be addressed in the future are:

- *Packaging* of complex devices (e.g., multisection lasers) requires special skills. In particular, a very careful package design is necessary for devices operating at very high speed.

- *Cost* will be a major issue. The complicated processing required leads to a low yield of working devices. This will have to be improved, and packaging costs will have to be reduced.

- *Optoelectronic integration* (integration of optical and electronic functions on the same chip) promises significant advantages. This will be a major area for research and development in the future.

CHAPTER 11. LIST OF SYMBOLS

The following list contains the *main* symbols used in the text, and their definitions. Some infrequently used symbols are not included. It has not been possible to avoid the use of the same symbol for different quantities, but the meaning should be clear from the text.
In some cases a numerical value is given in a bracket. These values apply to InGaAsP lasers for the 1.3 to 1.55 μm wavelength range. It must be stressed that the values are typical examples only and will depend on material, wavelength, laser dimensions, and laser structure.

a	gain slope, $a = dg_{act}/dN$ ($2.5 \cdot 10^{-20}$ m^2)
a'	change of refractive index with carrier density, $a' = dn_{act}/dN$
b	normalized propagation constant
B	recombination parameter
B	normalized bandwidth
c	velocity of light in vacuum
C	recombination parameter
C	feedback parameter
d	active layer thickness
D	fiber dispersion parameter
e	unit charge ($1.6 \cdot 10^{-19}$ As)
erfc	complementary error function
exp	exponential function
E	electrical field
f	frequency
f_{el}	electrical 3 dB frequency
f_B	data rate
f_m	modulation frequency
f_r	resonance frequency (5 GHz at a power of 5 mW)
$\bar{\bar{F}}$	transfer matrix
$F_s(t)$	photon number fluctuation
$F_s(\omega)$	spectral density of photon number fluctuation
$F_\phi(t)$	phase fluctuation
g	gain per unit length
g_{act}	gain per unit length in the active region, $g_{act} \approx a(N-N_0)$
g_{net}	net gain, $g_{net} = \Gamma\, g_{act}$
g_{th}	threshold gain, $g_{th} = \alpha_{int} + \alpha_{end}$
G'	gain per unit time, $G' = v_g\, g_{net} = v_g\, \Gamma\, g_{act}$
G_i'	gain per unit time for mode i
G_{th}'	threshold gain per unit time, $G_{th}' = v_g\, g_{th}$

$\hbar\omega$	photon energy; $\hbar$ is Planck's constant divided by 2π (0.8 eV for $\lambda = 1.55\ \mu m$)
$H(\omega_m)$	modulation transfer function
$Hel(\omega_m)$	electrical transfer function
I	current
$\bar{I}$	steady-state current
I_m	imaginary part
I_{th}	threshold current
j	imaginary unit
k	wave number, $k = 2\pi/\lambda$
K	proportionality between the damping rate and the square of the resonance frequency (0.3 ns)
K, K_t, K_z	linewidth correction factors
L	laser length (300 μm)
L	transmission distance
L_e	external cavity length
$L(\omega)$	line shape
m	mode number
M	order of periodic structure
M	modulation index
n	refractive index
$\bar{n}$	group index
n_{act}	refractive index of active region (3.5)
n_{eff}	effective refractive index (mode index)
n_{pas}	refractive index of passive layers (3.2)
n_{sp}	inversion parameter, $n_{sp} = N_{th}/(N_{th}-N_0)$ (2.5)
N	carrier density
$\mathcal{N}$	number of detected photons/bit
$\bar{N}$	steady-state value of carrier density
N_0	carrier density for transparency ($g_{act}=0$)
N_{th}	carrier density at threshold ($2\cdot10^{24}\ m^{-3}$)
prob	probability
P	power
P_D	detector equivalent noise power
P_e	error probability
$\bar{P}_i$	average power in mode i
P_L	local oscillator power
P_{out}	output power per facet
P_S	signal power
PSD	power spectral density
r	reflectivity (for the field) used with subscripts 0,1,2,3,e,B
R	facet reflectivity (for the intensity)
R	spontaneous emission rate, $R = v_g\, g_{th}\, n_{sp}$ ($10^{12}\ s^{-1}$)
R	wave amplitude in periodic structure
Re	real part
RIN	relative intensity noise per unit bandwidth
RIN_T	total relative intensity noise over a given bandwidth

R_1, R_2	facet reflectivities (for the intensity)
S	photon density ($4 \cdot 10^{21}$ m^{-3} at a power of 5 mW)
S	wave amplitude in periodic structure
S'	total number of photons, $S' = V\overline{S}$
$\overline{S}$	steady-state value of photon density
S_i	photon density in mode i
SNR	signal-to-noise ratio
t	time
T	bit period
v	normalized frequency
v_g	group velocity, $v_g = c / \overline{n}$ ($8 \cdot 10^7$ m/s)
V	active volume (10^{-16} m^3)
α	ratio between change in real and imaginary parts of the index (-5)
α'	loss per unit time, $\alpha' = v_g (\alpha_{int} + \alpha_{end})$
α_0	threshold gain (for the field) in a DFB laser
α_{end}	end loss (also called mirror loss, or facet loss), for a Fabry-Pérot laser $\alpha_{end} = \frac{1}{L} \ln\frac{1}{R}$ (for $R = R_1 = R_2$) (3000 m^{-1})
α_{int}	internal loss (3000 m^{-1})
β	spontaneous emission fraction
β	propagation constant, $\beta = n_{eff}\, k + j\, g_{net}/2$
β_0	Bragg propagation constant
γ	damping parameter, (8.5 GHz, at a power of 5 mW)
γ	DFB parameter, $\gamma = \sqrt{(\alpha_0 - j\delta)^2 + \kappa^2}$
Γ	confinement factor (0.3)
δ	deviation from Bragg propagation constant, $\delta = \beta - \beta_0$
$\delta(x)$	delta function
δf	frequency offset
δt	pulse width
$\delta\lambda$	spectral width
$\delta\nu$	linewidth
$\delta\nu_0$	linewidth of solitary laser
$\delta\nu_e$	linewidth of laser with external cavity
Δf	optical frequency variation
Δf	frequency range
Δg_{th}	threshold gain difference
$\Delta G'$	gain deviation from threshold, $\Delta G' = G'(N) - G_{th}'$
ΔI	current modulation amplitude
Δn	refractive index variation
ΔN	carrier density variation
ΔP	power variation
ΔS	photon density variation

$\Delta S_i'$	change in photon number due to one spontaneous emission event
ΔT	pulse width
$\Delta\lambda$	mode spacing
$\Delta\lambda$	wavelength variation
$\Delta\alpha_0$	threshold (field)gain difference
$\Delta\nu$	mode spacing
$\Delta\nu$	optical frequency variation
$\Delta\phi$	phase change
$\Delta\phi_i$	phase change due to one spontaneous emission event
$\Delta\omega$	angular frequency change
ε	nonlinear gain coefficient ($0.5 \cdot 10^{-23}$ m^3). Note that a number of authors define this parameter differently, corresponding to ε/Γ in the notation used here.
η	efficiency, used both for laser efficiency and detector efficiency
Θ	phase related to reflection
Θ_i	phase related to a spontaneous emission event
κ	coupling coefficient for periodic structure (DFB laser)
λ	wavelength
λ_0	Bragg wavelength for periodic structure
Λ	grating period
ν	optical frequency
τ	time interval
τ_d	round-trip time in the laser
τ_e	external cavity round-trip time
τ_{nr}	carrier lifetime due to nonradiative recombinations
τ_s	carrier lifetime-it is an approximation to take the lifetime as a constant since its value will depend on the carrier density (2 ns)
τ_s'	differential lifetime (at the threshold carrier density) (1 ns)
ϕ	optical phase
ω	frequency (angular)
ω'	corner frequency (angular)
ω_0	center optical frequency (angular)
ω_{el}	electrical 3 dB frequency (angular)
ω_{IF}	intermediate frequency (angular)
ω_L	local oscillator frequency (angular)
ω_m	modulation frequency (angular)
ω_r	resonance frequency (angular)
ω_S	signal frequency (angular)
ω_{th}	thermal cutoff frequency (angular)
$< >$	average value
$\| \ \|$	absolute value

CHAPTER 12. REFERENCES

1. H. C. Casey, Jr. and M. B. Panish, *Heterostructure Lasers*, Academic Press, Orlando (1978).

2. H. Kressel and J. K. Butler, *Semiconductor Lasers and Heterojunction LEDs*, Academic Press, Orlando (1978).

3. G. H. B. Thompson, *Physics of Semiconductor Laser Devices*, John Wiley & Sons, Chichester (1980).

4. G. P. Agrawal and N. K. Dutta, *Long-wavelength Semiconductor Lasers*, Van Nostrand Reinhold, New York (1986).

5. K. Petermann, *Laser Diode Modulation and Noise*, Kluwer Academic Publishers, Dordrecht (1988).

6. M. J. Adams, *An Introduction to Optical Waveguides*, John Wiley & Sons, Chichester (1981).

7. D. Botez, "Analytical approximation of the radiation confinement factor for the TE_0 mode of a double heterojunction laser," IEEE J. Quantum Electron., QE-14, 230-232 (1978).

8. K.-L. Chen and S. Wang, "An approximate expression for the effective refractive index in symmetric DH lasers," IEEE J. Quantum Electron., QE-19, 1354-1356, (1983).

9. C. H. Henry et al., "Partition fluctuations in nearly single-longitudinal-mode lasers," IEEE J. Lightwave Technol., LT-2, 209-216 (1984).

10. F. Koyama et al., "1.5-1.6 μm GaInAsP/InP dynamic-single-mode (DSM) lasers with distributed Bragg reflector," IEEE J. Quantum Electron., QE-19, 1042-1051 (1983).

11. T. E. Bell, "Single frequency semiconductor lasers", IEEE Spectrum, 20, 38-45 (1983).

12. W. Streifer et al., "Coupling coefficients for distributed feedback single- and double-heterostructure diode lasers," IEEE J. Quantum Electron., QE-11, 867-873 (1975).

13. J. Buus, "Mode selectivity in DFB lasers with cleaved facets," Electron. Lett., 21, 179-181 (1985).

14. R. S. Tucker, "High-speed modulation of semiconductor lasers," IEEE J. Lightwave Technol., LT-3, 1180-1192 (1985).

15. N. Ogasawara and R. Ito, "Longitudinal mode competition an asymmetric gain saturation in semiconductor injection lasers. I. Experiment," Jpn. J. Appl. Phys., 27, 607-614 (1988).

16. N. Ogasawara and R. Ito, "Longitudinal mode competition an asymmetric gain saturation in semiconductor injection lasers. II. Theory," Jpn. J. Appl. Phys., 27, 615-626 (1988).

17. R. Olshansky et al., "Frequency response of 1.3 μm InGaAsP high speed semiconductor lasers," IEEE J. Quantum Electron., QE-23, 1410-1418 (1987).

18. R. Olshansky et al., "Frequency response of an InGaAsP vapor phase regrown buried heterostructure laser with 18 GHz bandwidth," Appl. Phys. Lett., 49, 128-130 (1986).

19. K. Uomi et al., "Ultrahigh-speed 1.55 μm λ/4-shifted DFB PIQ-BH lasers with bandwidth of 17 GHz," Electron. Lett., 25, 668-669 (1989).

20. M. Osinski and J. Buus, "Linewidth enhancement factor in semiconductor lasers - an overview," IEEE J. Quantum Electron., QE-23, 9-29 (1987).

21. R. Schimpe et al., "Characterisation of frequency response of 1.5μm InGaAsP DFB laser diode and InGaAs PIN photodiode by heterodyne measurement technique," Electron. Lett., 22 453-454 (1986).

22. D. Marcuse, "Pulse distortion in single-mode fibers. 3: Chirped pulses," Appl. Opt., 20, 3573-3579 (1981).

23. C. H. Henry, "Theory of the linewidth of semiconductor lasers," IEEE J. Quantum Electron., QE-18, 259-264 (1982).

24. K. J. Vahala and A. Yariv, "Semiclassical theory of noise in semiconductor lasers - Part I," IEEE J. Quantum Electron., QE-19, 1096-1101 (1983).

25. K. J. Vahala and A. Yariv, "Semiclassical theory of noise in semiconductor lasers - Part II," IEEE J. Quantum Electron., QE-19, 1102-1109 (1983).

26. P. Spano et al., "Phase noise in semiconductor lasers: A theoretical approach," IEEE J. Quantum Electron., QE-19, 1195-1199 (1983).

27. C. H. Henry, "Theory of the phase noise and power spectrum of a single mode injection laser," IEEE J. Quantum Electron., QE-19, 1391-1397 (1983).

28. K. Kikuchi and T. Okoshi, "Measurement of FM noise, AM noise, and field spectra of 1.3 μm InGaAsP DFB lasers and determination of the linewidth enhancement factor," IEEE J. Quantum Electron., QE-21, 1814-1818 (1985).

29. K. Petermann, "Calculated spontaneous emission factor for double-heterostructure injection lasers with gain-induced waveguiding," IEEE J. Quantum Electron., QE-15, 566-570 (1979).

30. C. H. Henry, "Theory of spontaneous emission noise in open resonators and its application to lasers and optical amplifiers," IEEE J. Lightwave Technol., LT-4, 288-297 (1986).

31. K. Kojima and K. Kyuma, "Analysis of the linewidth of distributed feedback laser diodes using the Green's function method," Jpn. J. Appl. Phys., 27, L1721-L1723 (1988).

32. H. E. Rowe, *Signals and Noise in Communication Systems,* Chap. IV Sec. 3, D. Van Nostrand Company, Princeton (1965).

33. B. Daino et al., "Phase noise and spectral lineshape in semiconductor lasers," IEEE J. Quantum Electron., QE-19, 266-270 (1983).

34. L. E. Richter et al., "Linewidth determination from self-heterodyne measurements with subcoherence delay times," IEEE J. Quantum Electron., QE-22, 2070-2274 (1986).

35. S. Stein and J. J. Jones, *Modern Communication Principles*, McGraw Hill, New York (1967).

36. G. L. Abbas et al., "A dual-detector optical heterodyne receiver for local oscillator noise suppression," IEEE J. Lightwave Technol., 3, 1110-1122 (1985).

37. G. Jacobsen and I. Garrett, "Theory for heterodyne optical ASK receivers using square-law detection and postdetection filtering," IEE Proc. Part J, 134, 303-312 (1987).

38. I. Garrett and G. Jacobsen, "Theoretical analysis of heterodyne optical receivers for transmission systems using (semiconductor) lasers with nonnegligible linewidth," IEEE J. Lightwave Technol., LT-4, 323-334 (1986).

39. G. Ogita et al., "Linewidth reduction in DFB laser by detuning effect," Electron. Lett., 23, 393-394 (1987).

40. S. Takano et al., "Sub-MHz spectral linewidth in 1.5 μm separate-confinement-heterostructure (SCH) quantum-well DFB LDs," Electron. Lett., 25, 356-357 (1989).

41. M. Okai et al., "Narrow spectrum corrugation-pitch-modulated distributed feedback laser," OFC '90, paper ThE1 (1990).

42. H. Yamazaki et al, "250 kHz linewidth operation in long cavity 1.5 μm multiple quantum well DFB-LDs with reduced linewidth enhancement factor," OFC '90, paper PD33 (1990).

43. M. Okai et al., "Corrugation-pitch-modulated MQW-DFB laser with narrow spectral linewidth (170 kHz)," Photonics Technol. Lett., 2, 529-530 (1990).

44. R. W. Tkach and A. R. Chraplyvy, "Regimes of feedback effects in 1.5-μm distributed feedback lasers," IEEE J. Lightwave Technol. LT-4, 1655-1661 (1986).

45. J. Mørk et al., "Measurement and theory of mode hopping in external cavity lasers," Electron. Lett., 609-610 (1990).

46. D. Lenstra et al., "Coherence collapse in single-mode semiconductor lasers due to optical feedback," IEEE J. Quantum Electron., QE-21, 674-679 (1985).

47. J. Helms and K. Petermann, "A simple analytic expression for the stable operating range of laser diodes with optical feedback," IEEE J. Quantum Electron., QE-26, 833-836 (1990).

48. N. Schunk and K. Petermann, "Numerical analysis of the feedback regimes for a single-mode semiconductor laser with external feedback," IEEE J. Quantum Electron., QE-24, 1242-1247 (1988).

49. C. A. Park et al., "Single mode behaviour of a multimode 1.55 μm laser with a fibre grating external cavity," Electron. Lett., 22, 1132-1134 (1986).

50. S. Murata et al., "Spectral characteristics for 1.3 μm monolithic external cavity DFB lasers," Electron. Lett., 22, 1197-1198 (1986).

51. J. Buus, "Linewidth of monolithic external cavity DFB lasers," Electron. Lett., 24, 197-198 (1988).

52. S. Murata et al, "Over 720 GHz frequency tuning by a 1.5 μm DBR laser with phase and Bragg wavelength control regions," Electron. Lett. 23, 403-405 (1987).

53. C. F. J. Schanen et al., "Fabrication and lasing characteristics of λ = 1.56 μm tunable twin-guide (TTG) DFB lasers," IEE Proc. Part J, 137, 69-73 (1990).

54. S. Illek et al., "Over 7 nm (875 GHz) continuous wavelength tuning by tunable twin-guide (TTG) laser diode," Electron. Lett., 26, 46-47 (1990).

55. U. Koren et al., "Wavelength division multiplexing light source with integrated quantum well tunable lasers and optical amplifiers," Appl. Phys. Lett., 54, 2056-2058 (1989).

56. T. L. Koch et al., "GaInAs/GaInAsP multiple-quantum-well integrated heterodyne receiver," Electron. Lett., 25, 1621-1623 (1989).

About the Author

Jens Buus received the MSc and Ph.D. degrees at the Technical University of Denmark in 1976 and 1979, respectively. After spending four years as a postdoctoral research fellow in Denmark he joined Plessey Research Caswell Ltd. in 1983 as a senior principal scientist. He is presently senior chief physicist in the optoelectronics division at Caswell. Since 1976 Dr. Buus has published a large number of scientific papers dealing with different aspects of semiconductor lasers. He has served on a number of conference committees and is presently a member of the Technical Program Committee of the IEEE International Semiconductor Laser Conference.